£61.00

C000067025

INVESTIGATIONS OF THE TROPICAL ATLANTIC OCEAN

INVESTIGATIONS OF THE TROPICAL ATLANTIC OCEAN

Editor-in-Chief
V. N. Eremeev

Editorial Board
V. V. Efimov (Assistant Editor)
A. B. Polonsky
N. B. Bulgakov
G. S. Dvoryaninov
V. V. Knysh
G. K. Korotaev
N. A. Timofeev
N. B. Shapiro

Reviewer
V. I. Belyaev

VSP BV
P.O. Box 346
3700 AH Zeist
The Netherlands

© 1992 VSP BV

First published 1992

CIP-DATA KONINKLIJKE BIBLIOTHEEK, DEN HAAG

Investigations of the tropical Atlantic Ocean / ed.-in-chief V.N. Eremeev; editorial board V.V. Efimov; (ass. ed.) ... [et al.]; reviewer V.I. Belyaev; [transl. from the Russian]. - Utrecht: VSP
Transl. from: Issledovanie tropičeskoj Atlantiki: sbornik naučnych trudov. - Sevastopol:
Akademija Nauk USSR, Morskoj Gidrofizičeskij Institut, 1989.
ISBN 90-6764-143-X bound
NUGI 816
Subject heading: hydrophysical fields; Atlantic Ocean; seasonal variability / atmospheric phenomena / oceanic phenomena.

Typeset by TEV Ltd., Vilnius, Lithuania
Printed by Bookcraft (Bath) Ltd., Bath, UK

CONTENTS

PREFACE

Abstract — Discussed are the results of research conducted by the Marine Hydrophysical Institute of the Ukrainian SSR Academy of Sciences on the Soviet climatic RAZREZY program. On the basis of summarizing and modelling the *in situ* data, typical seasonal response of the Tropical Atlantic main hydrophysical fields to seasonal wind fluctuations is described. The kinematic structure of currents, heat and volume transport, and characteristics of their seasonal and synoptic variability, as recorded during the trans-Atlantic surveys in 1986–1988, are given. Spectral characteristics of the intratropical convergence zone variability in the Tropical Atlantic are provided.

It is common knowledge that tropical areas of the world's oceans represent an important source of climatic phenomena affecting the genesis of short-term climatic oscillations. As a result of interaction with the atmosphere, the tropical ocean accumulates heat, which is then transported to mid-latitudes. The study of air–sea interaction in the tropical region, heat accumulation and its transfer represent a major goal of climatic research on the RAZREZY program. The international TOGA (Tropical Ocean and Global Atmosphere) program dedicated to the study of tropical ocean regions and global weather anomalies also concentrates on the above problems.

Cruise research in the Tropical Atlantic energetic zone conducted by oceanographers of the Marine Hydrophysical Institute in the Amazon region during 1981–1984 involved regular seasonal CTD-surveying of the studied area. The main objective of these observations was to investigate the processes of heat transport from the Tropical Atlantic Ocean to the middle latitudes. It was assumed that thermal exchange between the Tropical and Sub-Tropical Atlantic regions is controlled by the system of jet currents in the western Tropical Atlantic Ocean constituted by the Guiana Current and the Antilles–Guiana Countercurrent. Results derived from the observational data allow conclusions as to the nature of circulation in the northwestern Tropical Atlantic and the latter's contribution to the meridional heat transport. The northwest-oriented boundary current constitutes part of the tropical anticyclonic gyre; during the greater part of the year it is turning northward, near 4–6°N, thereby giving rise to the North Equatorial Countercurrent (NECC). This pattern of currents is supported by field observations carried out by the Marine Hydrophysical Institute research vessels, as well as by numerical simulation. Computation of the tropical circulation, involving the treatment of historical hydrological data-bases also confirm the existence of this system of currents.

Considering the results, it becomes clear that an important role in the meridional heat transport must be attributed to the NECC which transports considerable volumes of heat eastward between 4 and 9° and then bifurcates in the eastern part of the Tropical Atlantic energetic zone. In connection with this, estimation of the meridional heat transport from the Tropical Atlantic Ocean toward the middle latitudes calls for an in-depth study of the entire system of currents in the Tropical Atlantic Ocean energetic zone rather than only that of its western section.

Since 1986, the research area in the Tropical Atlantic Ocean has been essentially extended, and it now encompasses the entire near-equatorial Tropical Atlantic zone (between 1.5°S and 12°N) stretching from the African continent to the South American coast. Research conducted during 1981–1988 have provided new evidence of the considerable seasonal variability of the hydrological characteristics in the Tropical Atlantic Ocean energetic zone and made possible description of their main peculiarities in the annual course.

Here we consider in detail the seasonal variability of the main hydrophysical and hydrometeorological characteristics of the Tropical Atlantic Ocean. Seasonal cycles of the fields of current velocity, temperature, and salinity, observed at long-term autonomous buoy stations, are discussed, along with the results of numerical modelling of the seasonal variability of circulation, temperature, and salinity fields in the Tropical Atlantic Ocean, involving a multi-layered nonlinear baroclinic model, which considers the basin's geometry and bottom topography configuration with the real external factors imposed (wind stress, ocean surface temperature, and salinity). The mechanisms for seasonal restructuring of the NECC are examined on the basis of theoretical analysis and observational data. Scrutinized are the air pressure fields, satellite-provided data on ocean surface temperature variability, the Intratropical Convergence Zone (ITCZ) meridional migrations, buoy data on temperature and currents, as well as on the trajectories of drifting buoys, and theoretical deductions. The most important problems relating to the study of large-scale perturbations in the Tropical Atlantic Ocean are outlined, the program of further investigations is given, and various methods and equipment are assessed.

The known inverse methods of calculating the currents field using probing data have allowed derivation of the currents' absolute velocities. The NECC kinematic structure is considered with involvement of the data from five large-scale hydrophysical surveys conducted in the Tropical Atlantic Ocean during 1986–1988. The NECC's spatial and seasonal peculiarities and determined qualitative characteristics of its location, depth, velocity, and mass transport have also been identified.

In general, this volume may be considered as a first stage in summarizing the RAZREZY investigations in the Tropical Atlantic Ocean, which is to be followed by a more scrupulous and thorough analysis of the whole hydrometeorological data-base.

Investigations of the Tropical
Atlantic Ocean, pp. 1 – 20
© VSP 1992.

Seasonal variability of the hydrophysical characteristics in the Tropical Atlantic Ocean: Part I. Materials and methods of data processing, noise level, and meteorological conditions

YU. V. ARTAMONOV and A. B. POLONSKY

Abstract — The data on the explorations conducted in the Tropical Atlantic Ocean are given. It is shown that the most complete CTD-data base and the results of currents observations are accumulated in the Marine Hydrophysical Institute (the Ukrainian SSR Academy of Sciences); the technique of data processing is described. The measurements provided by long-term autonomous buoy stations (surface moored buoys) are used to analyse the level of mesoscale and synoptic-scale noise and the kinetic energy density of synoptic-scale velocity fluctuations. Some characteristics of the seasonal variability of Tropical Atlantic Ocean meteorological parameters are given.

INTRODUCTION

The study of the seasonal variability of Tropical Atlantic Ocean hydrophysical characteristics constitutes one of the major goals of RAZREZY and FOCAL–SEQUAL oceanographic programs. The need for this problem to be resolved is coupled with the presence of the annual large-amplitude harmonic in the variance spectrum of Tropical Atlantic Ocean hydrophysical characteristics, which was revealed in the early 1980s. Thus, the identification and analysis of anomalous conditions accompanying the hydrophysical field evolution, as well as the determination of regularities in the low-frequency variability of the meridional heat transfer in the Tropical Atlantic Ocean, seem to be impossible without defining their normal seasonal course. On the other hand, investigations of the currents field's seasonal variations in surface and subsurface layers is of interest *per se* and seems of relevance from the viewpoint of practical implications.

The experimental data accumulated to date allow description of some general regularities in typical seasonal cyclicity of the hydrological characteristics. Seasonal cycle of the 20 °C isotherm's depth, thermocline characteristics, salinity maximum, as well as the heat content and dynamical topography are discussed

UDK 551.465

in refs. [1–16]. Surface currents in the Tropical Atlantic Ocean are studied using historical data on ship drifts [17, 18].

Two principal areas, where seasonal variability of the hydrophysical characteristics in the ocean's active layer is maximum were singled out: the first area is located in the northwestern section of the Tropical Atlantic Ocean near the annual mean position of the Intertropical Convergence Zone (ITCZ); the second area is observed in the Gulf of Guinea. The hydrophysical characteristics in these areas vary in counterphase with the annual cyclicity, with the borderline passing near 25°W. The other area of sharp phase fluctuations in the meridional direction is documented at the northern boundary of maximal seasonal temperature and salinity variations in the northwestern Tropical Atlantic Ocean. In addition, maximum seasonal variability of the North Equatorial Countercurrent (NECC) and of the South Equatorial Current (SEC) is registered in these areas. The general opinion is that there are two mechanisms behind this variability, namely, the oceanic response to trade winds and ITCZ seasonal variations and the local thermodynamic effects.

Notwithstanding the advances achieved in the study of Tropical Atlantic Ocean hydrophysical fields, the combined analysis, involving field-derived data, of the seasonal variability of kinematic and thermohaline structure of seawater, as well as of their relationship with the wind field, has not been as yet conducted. Intensive studies of the tropical zone carried out recently on the RAZREZY program, have essentially contributed to the hydrographic databases and instrument measurements, which made such analysis possible. Its results are submitted in this paper.

In Part I, the data and the methods of their processing are discussed, the high-frequency noise level and the variability of meteorological weather conditions at the ocean surface are analysed.

Part II is dedicated to the analysis of seasonal variations of temperature, salinity, and currents in the surface and subsurface ocean layers.

CHARACTERISTICS OF THE MATERIALS AND METHODS OF DATA PROCESSING

To characterize the seasonal variability of hydrophysical fields in the Tropical Atlantic Ocean, a number of climatic data-bases were used.

The data-base on current measurements includes approximately 500 autonomous buoy stations occupied between 20°S and 20°N up until 1987. These data are described in detail in refs. [19, 20]. The space–time distribution of data over the Tropical Atlantic Ocean is shown to be extremely nonuniform (Fig. 1). The majority of observations were carried out in the near-equatorial area in the vicinity of 23°30'W in 1974 (GATE project), in the northwestern Tropical Atlantic Ocean (RAZREZY program), and in the northeastern Tropical Atlantic Ocean (in the framework of the Soviet–Guinean Agreement). Duration of the observations ranged from several hours to dozens of days. The

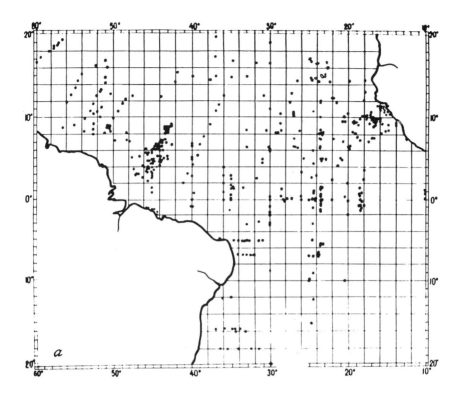

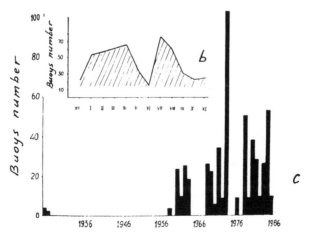

Figure 1. (a) distribution of autonomous buoy stations in the Tropical Atlantic Ocean occupied from 1926 to 1987; (b) annual changes in the number of buoy stations; (c) the number of buoy stations occupied in different years.

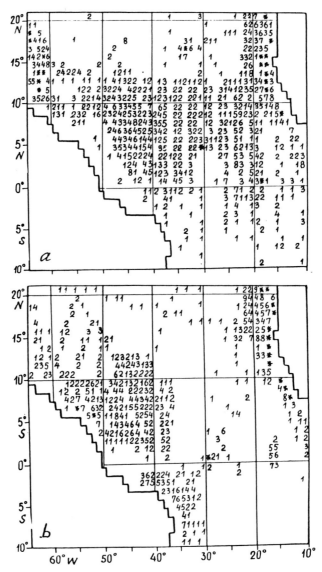

Figure 2. Climatic data support of the Marine Hydrophysical Institutes oceanographic data-base given in 1° × 1° squares (the sea surface: (a) in April; (b) in October).

majority of measurements were performed at 25-, 50-, 100-, and 200 m levels in winter and summer periods.

The hydrological data-base is being compiled by the Marine Hydrophysical Institute computer center [21]. It's core is constituted by the archieved information from the International Data Center on the Tropical and Subtropical Atlantic Ocean region bounded by 10°S and 15–65°W (about 30 000 hydrophysical stations); the data-base is being permanently updated with new information retrieved in the course of oceanographic surveys conducted by the research vessels of the Marine Hydrophysical Institute and of the State Oceanographic Institution Department in Sevastopol between 1°S and 12°N on the RAZREZY program. Until 1987, more than 3500 stations were occupied in this region, which has essentially contributed to the data-base on climatic fields (Fig. 2).

A detailed description of the hydrophysical data stored at Princeton University is given in ref. [22]. The support of these data with original observations from the North Tropical Atlantic Ocean is 2 to 3 times smaller than that of the hydrographic data-base compiled by the Marine Hydrophysical Institute (Fig. 3). Furthermore, a considerable part of the original data stored in the two files do not coincide, i.e. the files are based on the data-bases, which are factually independent of one another.

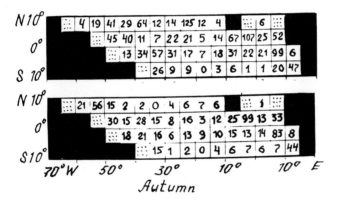

Figure 3. The number of hydrographic stations in the Princeton University data bank (in 5° × 5° squares [22]).

THE ERROR OF RECONSTRUCTION OF THE HYDROPHYSICAL FIELDS SEASONAL VARIABILITY CHARACTERISTICS

Two types of error add to the ultimate error of reconstruction of the normal hydrophysical fields' seasonal course: (i) the errors related to oceanographic instrumentation; and (ii) the errors occurring due to variability on the other scales, which in the case of discrete observations considerably distort seasonal characteristics. In addition, the second type of errors are distinctly prevalent [23].

To evaluate high-frequency fluctuations, basically, two types of data were used: long-term (15 days) buoy station observations, multiple hydrographic sounding data (at the Amazon test area, the GATE materials); and the data from recurrent surveys and hydrographic sections (the Amazon test area; standard section N 6 by the USSR Hydrometeorological Service in the northwest Tropical Atlantic and section along the 30°W).

The buoy station measurements provided the kinetic energy distribution by frequencies $(S(w))$, and allowed calculation of $\int_0^w S(w)\,dw$ and the ratio between the kinetic energy oscillations with diurnal cycle and the total kinetic energy. Calculations involved the data from 70 buoys deployed in various sections of the Tropical Atlantic Ocean, which gives a general notion about the space–time distribution of synoptic noise levels. The long-term hydrographic stations, occupied during the GATE and RAZREZY experiments, were invoked to calculate the total temperature and salinity dispersions at standard depths, the average daily dispersions, and the ratio (in %) between the daily fluctuations and the total ones, involving periods from several hours to dozens of days. Relying on the data from recurrent hydrologic surveys of sections and small test areas, we have calculated the space–time dispersions and expanded the thermohaline fields into empirical orthogonal functions.

Long-term observations were applied to calculate the current temperature spectra and the vector current velocity components, which permit to assess the intermittency of mesoscale fluctuations. Standard methods were used for this purpose [24].

Hydrological data from some sections were effectively used for calculations of one-dimensional spectra by the horizontal wavenumbers. Preliminarily, the trend was filtered out from the original series via the least-square technique, and the correlation function was then considered, whose form dictated the choice of model parameters. The obtained series was then, presented in the form of a sum of the process, governed by the autoregression (of order of p) and noise model. It was assumed that the 'useful' signal implies the occurrence of synoptic-scale wave phenomena, whereas 'noise' is conditioned by the high-frequency variability. These methods are discussed in detail in refs. [8, 23].

The data analysis indicates that a major contribution to the variability in the mesoscale part of spectrum is from semi-diurnal tidal oscillations. Diurnal tidal oscillations are visualized only in few cases. The typical amplitude of tidal velocity variations does not exceed $10\ \mathrm{cm\,s^{-1}}$, temperature 0.005–$1.5\ °\mathrm{C}$, and salinity 0.03–$1.0‰$. Their contribution to the total variability of the hydrophysical characteristics in various sections of the studied region attains 20–30%. However, it may account with depth for about 70–80%, for instance, over the seasonal thermocline, or at the interstitial waters levels (Figs. 4 and 5). The overall fluctuations dispersion then diminishes.

Thus, the average daily current velocity fluctuations do not really bear any impact upon the mean seasonal current characteristics, reconstructed from the direct measurements data. Their amplitude is typically small compared to that of the intensive jet flows seasonal variations*; and the fluctuations proper are filtered out by averaging over the observational period, which varies from one day to several dozens of days. Daily fluctuations of the hydrophysical fields greatly affect reconstruction of their typical seasonal cycle, as nearly

---

*See sections in Part II dedicated to the analysis of hydrophysical fields seasonal variability.

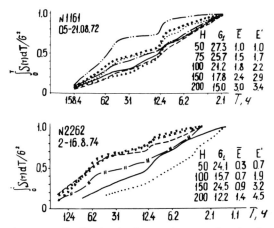

Figure 4. Kinetic energy distribution by frequencies at various levels normalized to the total dispersion: (a) 7°19′N and 21°51′W; (b) 6°38′N and 23°59′W; σ_1 is the standard deviation of diurnal variations, cm s^{-1}; $\overline{E}$ is the kinetic energy density in the 25 m layer, $\times 10^{-6}$ erg cm^{-2}, E' is the kinetic energy density of diurnal fluctuations, $\times 10^{-6}$ erg cm^{-2}; the dotted line indicates the 25 m level, the dashed line the 50 m level, the solid line denotes a 100 m level; triangulars denote the 150 m depth, and crosses the 200 m depth.

all data bases contain instantaneous original information (the typical sounding time is 1 h at most). Since typical dispersions of the daily and seasonal variations of thermohaline fields in the area of moderately intensive currents (e.g. NECC) are of the same order, this may lead to a considerable deformation of the seasonal response recovered from single measurements [8]. For high-frequency noise inhibition in the course of deriving average monthly and seasonal values, the data from several (in the areas with a large amplitude of the seasonal signal) to dozens (in the areas with weakly-marked seasonal variations) of independent soundings must be invoked. In this case, the error due to the high-frequency noise will be $\sqrt{n}$ times as small (where n is the number of independent observations). When 10–100 independent measurements (in the area of jet currents and weak large-scale flow, respectively) are used, the error due to high-frequency noise will be of the order of 10% . To attain such accuracy, different variants of space–time averaging of the original data were applied.

The filtering out procedure for synoptic-scale fluctuations of the hydrophysical fields represents a difficult task. In fact, daily fluctuations of temperature and salinity fields tend to intensify in the vicinity of jet streams. Concurrently, their contribution to the general dispersion increases.

The largest root-mean-square deviations of the currents' zonal component in the Tropical Atlantic Ocean, reaching several dozens of centimeters per second, are observed at the borderline between the NECC and the SEC, as well as between the Equatorial Undercurrent (EUC) and the SEC (Fig. 5). Here synoptic-scale temperature fluctuations in the thermocline attain 4–5 °C

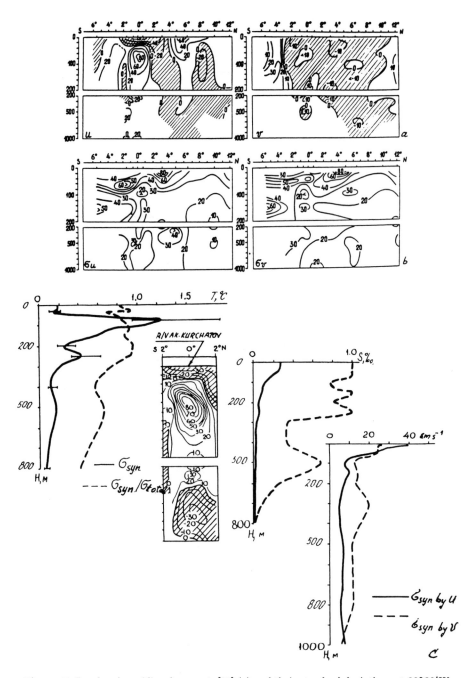

Figure 5. Zonal and meridional currents [25] (a) and their standard deviations at 23° 30′ W (b); vertical distribution of σ_{syn} and $\sigma_{syn}/\sigma_{sum}$ of temperature, salinity, and currents at the equator (R/V *Akademik Kurchatov*) (c).

and salinity variations 0.5‰. The total root-mean-square velocity fluctuations in the EUC main stream account for about 25% of the value of the mean flow. The root-mean-square deviation in the main stream of NECC is of the order of the mean velocity. An increase in fluctuation energy of the equatorial subsurface layer reflects the presence of equatorially-trapped waves generated by various sources.

The most spectacular dispersions of the currents' meridional component occur at the southern margin of the NECC, where the extreme values of standard deviations are several times larger than the mean current velocity.

It is possible that the instability of the zonal current system is responsible for the generation of the observed synoptic-scale current velocity fluctuations [26].

The vertical dispersion profile of the average daily velocity normally has a two-mode structure. The upper maximum occurs in the near-surface layers and is coupled with the strong synoptic variability of surface currents. The second extremum, particularly marked in the EUC area, is observed below the thermocline currents (Fig. 5).

The vertical profile of the ratio between the kinetic energy of synoptic-scale current velocity fluctuations and the total energy has also a two-mode structure. A specific feature for some of the buoy stations consists in the shifting of these maximums over the vertical versus the dispersion extremums of synoptic-scale velocity fluctuations. Similar peculiarities are observed in the temperature field.

With the purpose of determining regularities in the spatial distribution of the total energy, as well as of the synoptic oscillations in the seasonal cycle, all long-term buoy stations were grouped on a 2-month interval basis and by three meridional sections (35–45° as the western boundary layer, 20–25°W as the central part, and 10–20°W as the eastern boundary layer). Dispersions of current velocity fluctuations and the density of synoptic-scale fluctuations of the kinetic energy in various layers between 25 and 1500 m were calculated for every time interval and region.

The largest values of the dispersions and of the kinetic energy of synoptic fluctuations are observed during the summer/autumn period in the central and western sections of the equatorial zone, where these are in excess of $1000 \text{ cm}^2 \text{ s}^{-2}$ and $5 \times 10^6 \text{ erg cm}^{-2}$ (in the layer between 25 and 200 m), respectively. Away from the equator and in the eastern direction, synoptic fluctuations' intensity attenuates by approximately an order of magnitude. Also, the lower level of synoptic noise is documented during the winter/autumn period (Figs. 6 and 7).

Dispersion of synoptic fluctuations of the salinity field near the equator has its major minimum in the near-surface layers and is closely correlated with the synoptic variations of currents. It is less marked below the 200 m level and becomes negligibly small at a depth of 500–600 m. The second maximum is occasionally observed in the NEC and in the NECC at 300–500 m depths. The

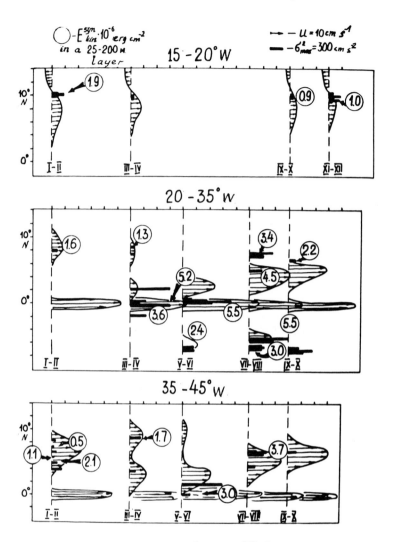

Figure 6. Spatial–temporal variability of σ^2_{max} and $\int_{25}^{200} \sigma^2 \, dz$ for three meridional bands: 15–20°; 20–35°; and 35–45° W.

contribution of synoptic salinity fluctuations to the total dispersion accounts for more than 50% down to the depth of 500–600 m.

Analysis of the results of expansion of the Tropical Atlantic Ocean thermohaline fields into empirical orthogonal functions has shown that the first empirical mode describes, on the average, about 80% of the total dispersion, and the initial two modes more than 90%. Moreover, the vertical statistical structure of these fields is stable both in the jet current area and beyond its bounds [8, 23]. The typical synoptic-scale wavelengths are 200–400 km. For their proper

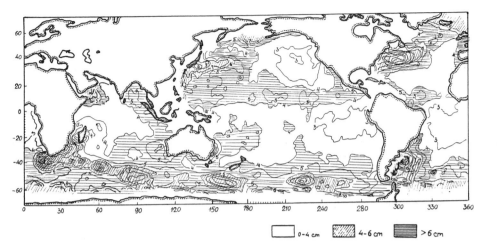

Figure 7. Spatial variations of σ^2_{syn} at the ocean level surface [33].

filtration, long-term (ranging from several dozens to several hundred days) observations of currents and thermohaline characteristics are needed. However, it must be admitted that such observations in the Tropical Atlantic Ocean remain for the most part unrivalled.

Let us evaluate the error in the reconstructed typical seasonal course of the hydrophysical fields, which occurs owing to diurnal variations. As the seasonal response in the Tropical Atlantic Ocean has basically the form of annual and semiannual variations [13, 23], we will expand into Fourier series the climatic fields at the grid joints. The total of the annual and semiannual harmonics will be considered as a net seasonal response, and the residual dispersion as a diurnal fluctuations manifestation. As a matter of fact, the latter dispersion is also partially induced by higher-frequency fluctuations and the deviation of hydrographic characteristics from those, described in terms of the typical seasonal cycle, due to the interannual variations. The first type of noise is analysed above. In the majority of squares adequately supported with observations, this noise does not exceed 10% of the amplitude of the seasonal response. The hydrographic field's interannual variations may attain 30–40% of the amplitude of the normal seasonal course [6, 13]. Therefore, the total error of single measurements is found to be equivalent, or even larger than the typical amplitude of the seasonal response.

In order to minimize the error of reconstruction of the hydrographic field's seasonal course parameters, temperature and salinity data were preliminarily smoothed and interpolated into $1° \times 1°$ grid joints, using the formula

$$\bar{f}_{i,j} = \sum_k |R_{x,y} - L_{i,j,k}| f_k \Big/ \sum_k |R_{x,y} - L_{i,j,k}|, \tag{1}$$

where $\bar{f}_{i,j}$ is the mean value at the grid joint, f_k is the observed value, $R_{x,y}$ is the effective radius ($R_x = 4°$, $R_y = 2°$), and $L_{i,j,k}$ is the space between the grid joints and the data point.

For some areas, adequately supported with observations, a simple arithmetic data averaging into regular grid joints (1° lat × 2° lon) was performed. A harmonic analysis then followed. This procedure permits the error of reconstruction of the typical seasonal cycle parameters to be reduced to 10–30%.

On the other hand, these parameters are similar to those used by Merle [13], and Arnault *et al.* [14], in the analysis of the hydrological fields' seasonal variations in the Tropical Atlantic Ocean, which makes possible an intercomparison of the acquired data. As the Princeton University data are smoothed and seasonally averaged, it is not possible for this procedure to be applied to these data. At the same time, the diagrams of dynamical topography based on these data-bases display mesoscale inhomogeneities conditioned by the spatial–temporal disagreement of observations [5]. The presence of high-frequency noise on scales of 200–300 km is also confirmed by one-dimensional spectra of the dynamical topography fluctuations. In order to smooth these inhomogeneities, the Princeton data were filtered out with the two-dimensional isotropic cosine Tjuky filter using a 9-point pattern

$$\bar{f}_{i,j} = 0.26 f_{i,j} + 0.13(f_{i,j+1} + f_{i+1,j} + f_{i,j-1} + f_{i-1,j}) \\ + 0.05(f_{i+1,j+1} + f_{i+1,j-1} + f_{i-1,j+1} + f_{i-1,j-1}). \tag{2}$$

The obtained density field was used for diagnostic calculations. A similar procedure was applied in ref. [27]. Polonsky *et al.* [28] have shown that as a result of smoothing the original fields from the Princeton University database, the calculated velocities and mass transport of large currents prove to be smaller by 30–50% as compared with the direct measurements data.

The accuracy of reconstructing parameters of the hydrographic field's seasonal course is achieved primarily by virtue of the sufficiently large volume of data collected in different years and phases of synoptic and mesoscale perturbations. Regretfully, the original *in situ* information on currents is less representative. The use of diurnal and long-term buoy stations, as has been pointed out above, reliably filter out the mesoscale noise. However, the latter's contribution to the error of reconstruction of the velocity field's typical seasonal cycle is not important. The total of the amplitudes of interannual velocity variations and of the diurnal fluctuations mentioned above, are larger (for some Tropical and Subtropical Atlantic Ocean areas, as well as for the T, S-characteristics) than the amplitude of the seasonal response (see also ref. [32]). With the scarce *in situ* observations, it is impossible to describe the kinematic structure and seasonal variability of the large currents in the Tropical Atlantic Ocean relying on the *in situ* data. This proved possible only for some areas in the northwestern and northeastern Tropical Atlantic Ocean and in the vicinity of 23°W. The error in the reconstructed large-scale kinematic pattern of the velocity field in some seasons is of the order of 30–50%. Therefore, we will analyse hereafter the data interpolated using equation (1), which were retrieved in the area bounded by the equator, 15°N, and 40–54°W (in the northwestern section of the Tropical Atlantic Ocean) and by 5–15°N and 15–24°W (in the northeastern section of the Tropical Atlantic Ocean). Some meridional sections adequately supported with observational data will also be handled. Incidentally, as distinct from the hydrographic data interpolation, in the majority of cases, the two-iteration procedure is applied which provides an additional smoothing of the original fields and the inhibition of noise of various origin [20].

In conclusion, we will briefly comment on the instrumental errors of various types of information discussed in the paper. The original information, constituting the groundwork for climatic data-bases, represents a set of experimental data compiled through the use of the following devices and techniques.

Bathythermograph observations with the vessel underway or drifting, whose error is ±0.2 °C. Seawater temperature measurements, using deep-water upsetting thermometers, yield an error of ±0.02 °C. More sophisticated probes of the ISTOK type provide better accuracy. The error alters, depending on the specific instrumentation involved, from 0.01 to 0.005 °C.

The error of salinity determination by the traditional method, i.e. through Knudsen–Oxner titration equals 0.02; the error of electrical conductivity measurements by modern probes, in terms of salinity reduces to 0.005‰.

The currents travelling in the direction of the vessel's drift, yield multiple random and regular errors which are hard to be estimated with relative accuracy. Wind-forced movement of the ship leads to the occurrence of a regular error in the surface current velocity. Single measurements of the velocity may yield a random error as large as 20 cm s^{-1}. The typical random error in the vessel's position, direction of movement and speed is evaluated to be ±2 km, ±1 °C, and ±15-20 cm s^{-1}, respectively. The error of mean velocity is essentially reduced when many observations are recorded; hence, for 100 observations, the standard error is reduced from 20 to 2 cm s^{-1} [24]. Over a large part of the Tropical Atlantic Ocean the typical density of the original information is 50 measurements in a 1° × 1° square. Therefore, the mean error of determination of the surface current velocity is normally 3–4 cm s^{-1}.

For a majority of current measurements in subsurface layers, the Alexeev BPV-2 recorder and MGI-1301 current meter (DISC) were used. Both instruments are designed to measure current velocity and direction separately; moreover, direction is registered instantaneously, whilst velocities are averaged over 200- and 4-s periods for the BPV-2 recorder and the DISC probe, respectively. The error of *in situ* current velocity measurements is roughly 1 cm s^{-1}, and for direction measurements is 1°. The errors in current measurements are related mainly to the measuring system's proper movement. Surface buoys are subjected to the effects of waves, wind, and currents and have their own trajectory of movement. This leads to considerable distortions of the velocity measurements, particularly in the high frequency part of spectrum. Alongside this, temporal averaging of the velocity modulus leads to the raising of mean velocity values, particularly in upper layers. This raising is considerable for slow currents where with velocities less than 10–20 cm s^{-1} it may attain 50%. In jet currents with velocities ranging from 100 to 150 cm s^{-1} this error is much smaller [29].

The error in computation of the dynamical heights [30], is proportional to the error of determination of the conventional specific volume (dV_t) and to the space (H) between the reference surface and the selected isobar: $dD = H\, dV_t$. Assuming the accuracy of determination of the conventional specific volume equal to ± 0.02 conv unit, this error, with H =3000 m, amounts to ± 60 dyn mm, and with H =1500 m, to ± 30 dyn mm. The relative error of dynamic height computations depends on the intensity of geostrophic circulation or on the magnitude of dynamical topography anomalies. In jet currents, this error does not exceed 15–30%. Large marine areas, where water circulation is less intensive, have a relative error in excess of 50%. With H =500–1000 m, the relative error in dynamic height computation reduces by 30–70%, therefore, in further analysis, reference surfaces up to the depth of 1000 m are involved.

Thus, the analysis of errors associated with *in situ* measurements of oceanographic characteristics indicates that these errors are, on the whole, considerably smaller than synoptic- and mesoscale variations of the hydrographic fields, which is consistent with the arguments presented in ref. [23].

SEASONAL VARIABILITY OF THE WIND FIELD AND HEAT FLUXES ON THE OCEAN SURFACE

The data of Hellerman and Rosenstein [31] on the Tropical Atlantic Ocean wind field were averaged over three meridional intervals (10–20°, 20–25°, and 35–45°W), which are characteristic of the conditions in the eastern, central, and western sections of the Tropical Atlantic Ocean.

The wind field over the Tropical Atlantic Ocean displays a strongly marked annual cycle with a distinct borderline between the northeasterly and southeasterly trade winds. This frontier is observed in the form of zero values of the tangential wind stress meridional component. The zonal component minimum also coincides with the ITCZ position. In the eastern Tropical Atlantic Ocean, the Intratropical convergence zone is located farther to the north throughout the year than in the central and western parts. The range of the ITCZ's seasonal variations constitutes about 10° latitude. It occupies its northernmost position in the eastern Tropical Atlantic Ocean at the end of summer and beginning of autumn (August, September; 12°N), and the southernmost position (below the equator) in the western Tropical Atlantic Ocean during the second half of winter and spring (January–April) (Figs. 8 and 9).

To the east (10–20°W), between the equator and 15°N, the tangential wind stress zonal component remains positive almost throughout the year and attains its maximum in July–September ($\tau_{0x} > 0.2$ dyn cm^{-2}). An increase of τ_{0x} during July–August is accompanied by an intensification of the meridional

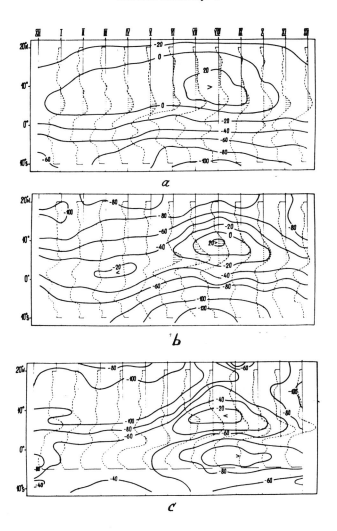

Figure 8. Zonal components of currents (u) and tangential wind stress (τ_{0x}) at the ocean surface (solid lines correspond to $\tau_{0x} = 10^2$ dyn cm^{-2}).

component τ_{0y} owing to the development of the southwesterly trade wind. South of the equator, an easterly wind (the southeasterly trade wind) is observed throughout the year, being at its maximum during June–September ($|\tau_{0x}| > 1$ dyn cm^{-2}). Approximately at same time, the meridional component τ_{0y} reaches its maximum (> 0.6 dyn cm^{-2}, Fig. 8).

Trade winds in the central Tropical Atlantic Ocean (25–30°W) occur mainly between 5 and 10°N during July–September. On both sides of this zone, easterly winds (northerly and southeasterly trade winds) prevail during the year. In September–July the intermediate area is characterized by the minimal zonal

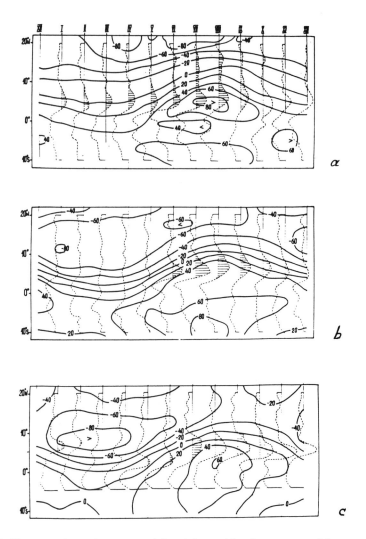

Figure 9. The current's zonal component (u) and the meridional component of the tangential wind stress (τ_{0y}) at the ocean surface (solid lines indicate $\tau_{0y} = 10^2$ dyn cm^{-2}).

component ($\tau_{0x} \sim 0.2$ dyn cm^{-2}). The sign of the tangential wind stress meridional component τ_{0y} alternates along this borderline. The northeasterly trade wind is observed to be most intensive in December–January, and the southeasterly trade wind in July–August (Figs. 8 and 9).

In the western part of the tropical zone (35–45°W) the wind is west-oriented. During the first half of the year, the northeasterly trade wind is migrating toward the equator, and minimum τ_{0x} occurs in the southern hemisphere. From April to August minimum τ_{0x} is shifting northward, and in August it is to be sought in the area between 5 and 10°N. The southeasterly trade wind is intensifying from July to January, with maximum τ_{0x} being documented in August–October (1 dyn cm^{-2}; Fig. 8).

The meridional component of the northeasterly trade wind reaches its maximum in January–April ($|\tau_{0y}| > 0.8$ dyn cm^{-2}); in the southeasterly trade wind maximum τ_{0y} (> 0.6 dyn cm^{-2}) occurs in July–September, and the largest amplitudes of the annual harmonic of τ_{0x} and τ_{0y} are observed within the ITCZ. Moreover, in the field of τ_{0x}, these attain a maximum in the western Tropical Atlantic Ocean (> 0.4 dyn cm^{-2}). The amplitude of variations of τ_{0y} in the eastern Tropical Atlantic Ocean is close to the amplitude of the annual harmonic of τ_{0y} in the western Tropical Atlantic Ocean and is larger than 0.6 dyn cm^{-2}. In the area of the North Equatorial Countercurrent, vorticity in the wind field attains the magnitude of maximum seasonal variations as indicated earlier in Katz and Garzoli [11], Henin and Hizard [32]. The amplitude of the seasonal wind field variability in the western Equatorial Atlantic Ocean (between 5–10°N and 40–50°W) is larger than a similar amplitude for the whole of the remaining Tropical and Subtropical Atlantic Ocean. These wind field fluctuations generate an oceanic response, which has a form of a high-amplitude seasonal variability of drift currents and subsurface hydrophysical characteristics in the vicinity of the North Equatorial Countercurrent. It will be appropriate to note the presence of a strongly marked peculiarity of air–sea interaction in this region: the mean volume of heat transferred into the atmosphere annually through the ocean surface reaches about 10^{14}W, with the major heat exchange occurring in December–February when the northeasterly trade wind is strongest (which facilitates powerful latent heat fluxes), and the radiation heating is minimal (Fig. 10).

North of 10°N the amplitude of seasonal radiation balance variations increases, and notwithstanding the degrading of the seasonal variability of the wind field, seasonal fluctuations of heat fluxes at the ocean surface become more enhanced. At the northern boundary of the tropical zone their amplitudes are as large as 100–150 W m^{-2}.

The rest of the Tropical Atlantic Ocean annually receives, on average, about 25 W m^{-2}, and the eastern section (in the African upwelling zone) in excess of 50 W m^{-2}.

Thus, the wind field and heat exchange between the air and ocean in the Tropical Atlantic are characterized by the following features:

(1) The annual minimum tangential wind stress zonal component in the vicinity of the ITCZ axis (where $\tau_{0y} \approx 0$) is located approximately at 5°N and equals about ~ 0.2 dyn cm^{-2}.

(2) North of 13° N, the mean annual tangential wind stress monotonically decreases toward the central part of the Subtropical gyre.

(3) Seasonal fluctuations of the tangential wind stress are greatest in the ITCZ area, where the amplitude of the annual harmonic for τ_{0x} and τ_{0y} is ~ 0.5 dyn cm^{-2}.

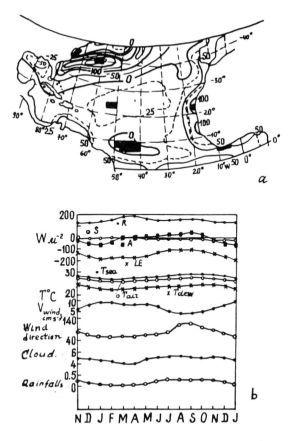

Figure 10. Average annual heat fluxes at the ocean surface ($W m^{-2}$) (a) and the seasonal course of meteorological parameters between 9–10° N and 40–50° W 32: A indicates the total heat fluxes at the surface; R is the radiation balance; and LE are the latent heat fluxes [32].

(4) The annual harmonic amplitude for τ_{0x} decreases from west to east by 2 to 3 times, whereas for τ_{0y} it remains roughly constant.*

(5) The amplitude of semiannual variations of τ_{0x} and τ_{0y} in the vicinity of the ITCZ equals 0.1–0.2 $dyn cm^{-2}$.*

(6) The Tropical Atlantic Ocean absorbs annually, on average, 25–50 $W m^{-2}$, with the major amount (in excess of 50 $W m^{-2}$) being transferred in the West African upwelling zone.

(7) An area is identified between 5–10° N and 40–50° W, where heat emission through the ocean surface takes place, attaining annually about 10^{14} W.

(8) The largest amplitudes of seasonal variations of the heat flux (100–150 $W m^{-2}$) are observed in the North Tropical Atlantic Ocean.

---

*See Figs. 13 and 14 in Part II.

REFERENCES

1. Artamonov, Yu. V. Seasonal variability of the major hydrophysical characteristics in the Tropical Atlantic. Abstract of the thesis. Moscow: (1989), 23 p.

2. Artamonov, Yu. V., Bulgakov, N. P. and Cheremin, V. N. Subtropical waters in the equatorial countercurrents of the Atlantic Ocean. Moscow: Dep. VINITI, N 9048-B87 (1987), 16 p.

3. Artamonov, Yu. V., Bulgakov, N. P. and Lvov, V. V. The kinematic and thermohaline structure of seawater in the area of genesis of the North Equatorial Countercurrent. Moscow: Dep. VINITI, N 3743-B88 (1988), 25 p.

4. Artamonov, Yu. V., Bulgakov, N. P., Surikova, I. A. and Urikova, N. V. Seasonal variability of Tropical Atlantic thermohaline fields. Moscow: Dep. VINITI, N 4132-B88 (1988), 29 p.

5. Artamonov, Yu. V., Polonsky, A. B. and Pereyaslavsky, M. G. Investigations of the large-scale circulation in the northeastern section of the Tropical Atlantic. Moscow: Dep. VINITI, N 992-B87 (1987), 25 p.

6. Bubnov, V. A. and Navrotskaya, S. E. Variability of the dynamical topography and surface geostrophic currents in the central Tropical Atlantic. *Okeonologia* (1985) **25**, 395–402.

7. Bubnov, V. A. Circulation in the equatorial zone of the World's ocean. Doctoral dissertation. Moscow: (1986), 530 p.

8. Bulgakov, N. P. and Polonsky, A. B. The main hydrophysical fields in the Tropical Atlantic Ocean and their variability. Sevastopol: Preprint MHI (1985), 44 p.

9. Khanaichenko, N. K. *The System of Equatorial Countercurrents in the Ocean.* Leningrad: Gidrometeoizdat (1974), 176 p.

10. Khlystov, N. Z. *The Structure and Dynamics of Waters in the Tropical Atlantic.* Kiev: Nauk. dumka (1976), 164 p.

11. Garzoli, S. and Katz, J. L. The forced annual reversal of the North Equatorial Countercurrent. *J. Phys. Oceanogr.* (1983) **13**, 2082–2090.

12. Houghton, R. W. Seasonal variations of the subsurface thermal structure in the Gulf of Guinea. *J. Phys. Oceanogr.* (1983) **13**, 2070–2081.

13. Merle, J., Fieux, M. and Hizard, P. Annual signal and interannual anomalies of sea surface temperature in the eastern Equatorial Atlantic Ocean. *GATE Symp. Oceanogr. and Surface Layer Mean Role.* Kiel. Oxford. C. A. (1980) **2**, 77–101.

14. Merle, J. and Arnault, S. Seasonal variability of the surface dynamic topography in the Tropical Atlantic. *J. Mar. Res.* (1985) **43**, 267–288.

15. Cane, M. A. and Houghton, R. W. Atlantic seasonality: Observations. Further Progress in Equatorial Oceanography. A report of the US TOGA. Workshop on the dynamics of the equatorial ocean. Honolulu H. I. August 11–15, 1986. Fla. (1987), 215–233.

16. Hastenrath, S. and Merle, J. Annual cycle of subsurface thermal structure in the Tropical Atlantic Ocean. *J. Phys. Oceanogr.* (1987) **17**, 1518–1538.

17. Richardson, P. L. and McKee, T. K. Average seasonal variation of the Atlantic equatorial currents on historical ship drifts. *J. Phys. Oceanogr.* (1984) **14**, 1226–1238.

18. Richardson, P. L. and Walsh, O. Mapping climatological seasonal of surface currents in the Tropical Atlantic using ship drifts. *J. Geophys. Res.* (1986) **91** (C9), 10,537-10,550.

19. Artamonov, Yu. V., Bulgakov, N. P., Belous, L. M. *et al.* The main currents in the Tropical Atlantic as recovered from *in situ* measurements. Moscow: Dep. VINITI, N 9047-B87 (1987), 28 p.

20. Polonsky, A. B. On an average circulation of the Equatorial Atlantic Ocean. *TO-AN* (1987), 10–12.

21. Chernova, A. V. and Chmutov, M. V. The system of storage and treatment of oceanographic data using the OKA data control system. Automated systems for collecting and processing oceanographic data. Sevastopol: Preprint MHI (1987), 30 p.

22. Levitus, S. and Oort, A. Global analysis of oceanographic data. *Bull. Am. Met. Soc.* (1977) **38**, 1270–1284.

23. Belevich, R. R., Bulgakov, N. P., Polonsky, A. B. and Popov, Yu. P. Variability of the hydrophysical fields in the Equatorial Tropical Atlantic. *Tr. GOIN* (1985) **173**, 66–79.

24. Rozhkov, V. A. *The Methods of Probability as Applied to the Analysis of Oceanographic Phenomena.* Leningrad: Gidrometeoizdat (1979), 279 p.

25. Bubnov, V. A. *et al.* Graphical presentation of the USSR oceanographic observations in the Tropical Atlantic during GATE (June to September 1974) PSMAS/U of M. Technical Report N TR79-1, Miami (1974), 163 p.

26. Philander, S. G. H. and Duing, W. The oceanic circulation of the Tropical Atlantic and its variability as observed during GATE. *Deep-Sea Res., GATE Suppl.* (1980) **2**, 1–27.

27. Sarkisyan, A. S. *The Methods and Results of Computations of the World's Ocean Circulation.* Leningrad: Gidrometeoizdat (1986) 152 p.

28. Polonsky, A. B., Goryachkin, Yu. N., Kazakov, S. I. and Pereyaslavsky, M. G. Large-scale circulation in the northwestern Tropical Atlantic and its seasonal variability. Experimental studies in the Tropical Atlantic. Moscow: Dep. VINITI, N 4986-B85 (1985), 26–37.

29. Halpern, D. Review of intercomparisons of moored current measurements. Oceans-77 Conference record. V. Z. Los Angeles, Bonaventure, October 17–19 (1977), 32–47.

30. Fomin, L. I. *Theoretical Fundamentals of the Dynamical Method and its Application to Oceanology.* Moscow: Izd. AN SSSR (1961) 192 p.

31. Hellerman, S. and Rosenstein, M. Normal monthly wind stress over the World Ocean with error estimates. *J. Phys. Oceanogr.* (1983) **13**, 1093–1104.

32. Henin, C. and Hizard, P. The North Equatorial Countercurrent observed during the FOCAL in the Atlantic Ocean. July, 1982 to August, 1984. *J. Geoph. Res.* (1987) **92** (C4), 3751–3758.

33. Cheney, R.E., Margh, J. C. and Beckley, Br. D. Global mesoscale variability from collinear tracks of SEASAT altimeter data. *J. Geoph. Res.* (1983) **88** (C7), 4343–4354.

34. Banker, A. F. Computation of surface energy flux and annual air-sea interaction cycle of the North Atlantic Ocean. *Mon. Weath. Rev.* (1976) **104**, 1122–1138.

*Investigations of the Tropical
Atlantic Ocean*, pp. 21 – 42
© VSP 1992.

Seasonal variability of the hydrophysical characteristics in the Tropical Atlantic Ocean: Part II. Currents and the *T*, *S*-characteristics

YU. V. ARTAMONOV and A. B. POLONSKY

Abstract — Climatic data discussed in Part I are used to describe the typical seasonal cycle of hydrophysical fields in the Tropical Atlantic Ocean. We mainly concentrate on the latitudinal section bounded between the equator and 12°N, which is best supported with relevant observations. The major phases of the genesis, evolution, and degradation of the North Equatorial Countercurrent are considered. The amplitudes of annual and semi-annual variations of the basic hydrophysical fields and phase fluctuations between seasonal variations of meteorological and oceanographic parameters are presented.

INTRODUCTION

It was shown in Part I of this study [1] that meteorologic conditions over the Atlantic Ocean surface undergo large-amplitude seasonal fluctuations. These are known to produce an oceanic response, which, notwithstanding the high level of high-frequency noise, is distinctly traceable in the annual and semi-annual harmonics [2–7]. Part II of the study is dedicated primarily to the analysis of the hydrophysical characteristics' seasonal variability in surface and subsurface layers relying on the contemporary climatic data-bases described in detail in ref. [1].

SEASONAL VARIABILITY OF TEMPERATURE, SALINITY AND CURRENTS

Surface currents in the Tropical Atlantic Ocean are predominantly of a zonal nature [8–10]. Therefore, we will restrict ourselves to the analysis of seasonal variability of the zonal velocity component of the North Equatorial Current, South Equatorial Current, and the North Equatorial Countercurrent.

The specific feature of ocean surface temperature distribution is associated with the presence of an extensive frontal zone, which, during existence of the Equatorial Countercurrent, is located in the vicinity of its northern boundary over the entire Tropical Atlantic Ocean (the subtropical equatorial front)

UDK 551.465

(Fig. 1). The subtropical equatorial front and its seasonal fluctuations are best pronounced on the ocean surface in the eastern section, where the depth of the upper quasi-isothermal layer is minimal. During the July–August period, the front and the Equatorial Countercurrent's northern boundary occupy the northernmost position (20°N), and in February–April the southernmost position (below 10°N). It will be appropriate to emphasize here that there are two branches of the Equatorial Countercurrent to be readily visualized in the surface layer during summertime. The northern branch reaches 18°N, which is consistent with the data recovered from diagnostic computations, while the southern branch passes at 4–5°N and gives rise to the Guinean Current. North of the front, the temperature minimum (16 °C) occurs in the eastern Tropical Atlantic in February; it is coupled with the development of an intensive coastal upwelling near north-west Africa. South of the front, there is a thermal equator, in the vicinity of which minimal dispersion of mean monthly temperatures is documented. In the eastern section of the region, the thermal equator roughly coincides, as the annual data averages indicate, with the North Equatorial Countercurrent's main stream, and its seasonal course reflects spatial fluctuations of this countercurrent.

South of the thermal equator seawater temperature declines. The largest meridional temperature gradients occur in July–September. The temperature minimum (24 °C) is found south of the equator in the area of west-oriented currents with minimal velocity. This temperature minimum is related to the equatorial divergence. The maximum amplitude of annual velocity oscillations (20 m s^{-1}) is located at 1–2°N.

During the first half of the year, the subequatorial front in the central Tropical Atlantic Ocean is located between 5 and 10°N. The North Equatorial Countercurrent in the form of a surface flow oriented eastward does not exist at this time of the year, and the entire region between 20°N and 10°S is involved in the transport of the North Equatorial Current and South Equatorial Countercurrent in the westerly direction. The border between these two currents is identified by the local minimum of the west velocity component. The southern periphery of the subequatorial front, which separates the South Equatorial Current warm waters from cold waters of the North Equatorial Current during this time of the year, passes approximately along this boundary.

During the second half of the year, i.e. from May to September, the Subequatorial front shifts to the north and occupies in October the area between 15 and 20°N. An east-oriented stream (the North Equatorial Countercurrent) is generated south of the front in June.

The Subequatorial front in the western section of the Tropical Atlantic Ocean is weakly marked in the ocean surface layer, as compared to the eastern and central sections, due to the small thickness of the upper mixed layer. From February to May, the front is closest to the geographic equator, and the thermal equator proves to be nearly completely forced out into the southern hemisphere. From June to September, the Subequatorial front is shifting northward, and

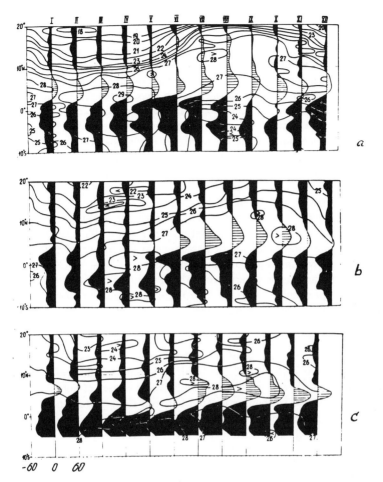

Figure 1. Surface temperature and current distributions averaged between 10 and 20°W (a), 25 and 30°W (b), and 35 and 45°W (c).

the thermal equator is tracked between the equator and 5°N; the North Equatorial Countercurrent in the form of an easterly flow is observable at the ocean surface from May to February. In March through April, the North Equatorial Countercurrent manifests itself in the form of local minima of west current velocity components. Note that from March to June actually two branches of the countercurrent exist: the southern branch which evolves in May–June; and the northern branch which represents the degradating Equatorial Countercurrent. The double-jet structure of the Equatorial Countercurrent in the west Tropical Atlantic Ocean was initially predicted following the numerical experiment in ref. [5]. It will be noted that the generating mechanisms behind these two flows of the Equatorial Countercurrent in the eastern and western Tropical Atlantic Ocean are thoroughly different. Bifurcation of the Equatorial Countercurrent

in the eastern section is traceable throughout the year due to the specific geometry of the coast and bottom topography. It is better marked during the
summer–autumn period, with the Equatorial Countercurrent being intensive
due to the intensification of the southwesterly trade winds. The double-stream
pattern of the countercurrent is best observed during spring, which is explained
by the difference in the rate of the ocean's response to the displacement of the
Intratropical Convergence Zone to the north or south from its middle position.

The largest amplitudes of annual fluctuations of the NECC zonal component
($\sim 30 \ \mathrm{cm \ s^{-1}}$) are observed in the northwestern Tropical Atlantic Ocean, where
these are half as large as the amplitudes in the central and eastern sections of
the region. Conversely, the amplitudes of annual thermal oscillations in the
western area are smaller than in the central and eastern ones (1–1.5; 2–3; 4–
5 °C, respectively, within the Subequatorial front region). In the vicinity of
the thermal equator, where seasonal variability of currents is considerable, the
annual sea surface temperature variations are modest. This is accounted for by
the fact that formation of the heat balance in the upper navigational layer is
influenced by two factors: the meridional shifting of the NECC–Subequatorial
front system and seasonal variance of thermal fluxes at the ocean surface. This
explains, in particular, why the thermal equator in the western and central
Tropical Atlantic Ocean shifts in the course of the year to the south from the
North Equatorial Countercurrent's main stream.

Salinity distribution in the ocean surface layer seems to be more complicated.
High salinity, characteristic of the central areas of the north and south anticyclonic gyres, is observed in the north and south of the region. The weakly
pronounced frontal zone partitions highly-saline waters of northern origin from
the area of relative freshening located between the equator and 5–15°N in the
ITCZ area. Seawater freshening in the ITCZ area is enhanced due to the
advection of waters of low salinity affected by river run-offs.

From March to September, the salinity front in the eastern Tropical Atlantic
Ocean is encompassed by the NECC; and from October to January it occupies
the northernmost position (15–18°N) and shifts from the NECC area into the
region where westerly transport by the Canary current occurs. The highest
degree of freshening occurs between 6 and 12°N and persists from September
to January. The largest amplitude of annual salinity variations ($\sim 2\text{‰}$) is
observed near 6°N. Freshening of seawater in the central part of the region
between 10°N and the equator occurs largely due to rainfall in the ITCZ; the
effects of the continental run-off here is not traceable. The amplitude of annual
salinity variations does not exceed 1‰. Salinity concentration in the western
Tropical Atlantic Ocean, more exactly, at the border of the region under study
amounts to the values which are characteristic of the central part of the North
subtropical anticyclonic gyre (37.2‰). An intensive freshening of seawater
induced by the transport of the Amazon and Orinoco waters to the east by
the Guiana Current–NECC system, takes place in the area between 10–15°N
and the equator from July to November. The amplitude of the annual salinity
course at 6°N attains 2‰. Another, less intensive seawater freshening occurs

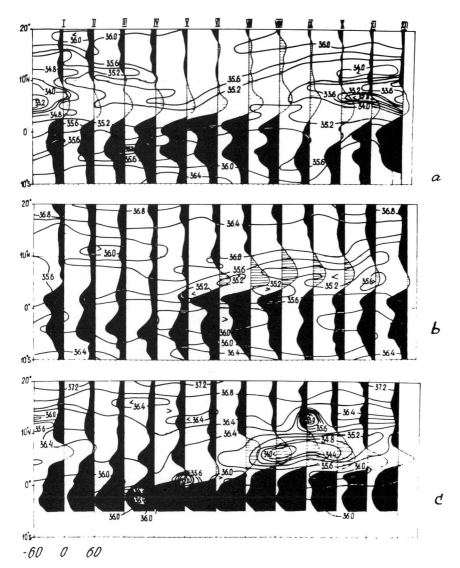

Figure 2. Surface salinity and currents distributions averaged between 10 and 20°W (a), 25 and 30°W (b), and 35 and 45°W (c).

south of the equator from March to May. It is induced by heavy rainfall in the ITCZ, which is located at this time of the year near the equator.

More detailed analysis of the salinity field in the NECC area shows that low salinity related to continental run-off occurs west of 50°W almost throughout the year. When the NECC vanishes from the sea surface between 20 and 50°W, salinity remains relatively high from January to May. From May to December, when the NECC outcrops, seawater of low salinity is transported from coastal areas to the open ocean. From August to November, these waters reach 35°W.

Near the eastern coast, the freshening period starts in January and terminates in December, with a tongue of low salinity waters extending in September–November as far as 30°W. In the eastern half of the Atlantic Ocean, waters of low salinity propagating in a westward direction are traceable far off the coast. Formation of these large-scale ocean regions with low salinity is associated in this case with heavy rainfall in the ITCZ rather than with the advection [2].

Two periods of seawater freshening have been observed to take place in the western part of the region. The first period is from January to May and is related to the continental run-off and heavy rainfall within the ITCZ, which at this time of year is located near the equator. The freshening area is stretched eastward almost up to 30°W. The second period of salinity lowering is observed in August–October and is prompted by an increase in run-off with the northeastern trade wind being attennuated. The area of freshening does not extend far to the east along the equator, as seawater of low salinity is transported by the Guiana Current in a northwestern direction. The freshening process in the eastern equatorial section is observed from December to May and is conditioned by the transport of low salinity waters from the Gulf of Guinea by the South Equatorial Current. Thus, two major factors control the formation of large areas of seawater freshening in the Tropical Atlantic Ocean, namely, the advection of seawater of low salinity transported from coastal areas and rainfall. The tongue of low salinity water in the western part of the North Equatorial Countercurrent is produced mainly by the Amazon water advection to the open ocean, whereas in the eastern region the major mechanism responsible for freshening is coupled with ITCZ rainfall. The pattern is reversed in the equatorial region: the largest freshening in the western area is produced by ITCZ rainfall, and in the eastern area by the transport of the Gulf of Guinea waters of low salinity by the South Equatorial Current.

Semi-annual temperature, salinity and currents velocity fluctuations stand out in a most emphatic fashion in the eastern Tropical Atlantic Ocean. The maximum amplitude of the semi-annual harmonic of temperature fluctuations (1.5 °C) occurs between 5 and 10°N, that of salinity variations (0.5‰) between 3 and 7°N, and that of current velocities (20 cm s^{-1}) at the equator. In the western Tropical Atlantic Ocean semi-annual fluctuations are most pronounced in the current's field. Here, two areas with large amplitudes of semi-annual current fluctuations are registered. The first area is located in the vicinity of the NECC's mean annual position at 6–8°N. The second maximum is observed near 2–4°N and is related to semi-annual variations of the Guiana Current, which are a result of the complicated superposition of annual harmonics in the variability of gradient and drift current velocity component shifted in phase from each other by a period of several months.

SEASONAL VARIABILITY OF THE CURRENTS' FIELD IN THE SUBSURFACE LAYER

Water circulation as recovered from in situ *measurements*

The most complete synopsis of *in situ* observations of Tropical Atlantic Ocean currents is given in refs. [3, 10, 11]. Circulation in the subsurface layer of the eastern section of the Tropical Atlantic Ocean and the kinematic structure of currents at 23°30'W during the GATE investigations is given in these studies. Several papers are also published where episodical observations are examined [8–10]. Regretfully, no fundamental studies have been as yet published, save some preliminary works [12, 13], which would have handled the data covering the entire Tropical Atlantic Ocean. Therefore, we will attempt to consider in finer detail the kinematic structure and the mass transport of currents and countercurrents in the Tropical Atlantic Ocean using *in situ* observations. We will focus mainly on the system of the equatorial countercurrents, which are best supported with the data.

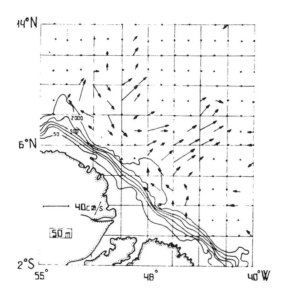

Figure 3. Annual averages of currents based on *in situ* measurements at 50 m depth.

It is readily visualized from the annual maps of currents and sections in the northwestern Tropical Atlantic Ocean that the NECC is generated west of 45°W between 2–4° and 6–8°N. In addition, it develops mainly in the upper 300 m layer (Figs. 3 and 5a). Between 30 and 40°W, it forms one east-oriented flow with the Equatorial Undercurrent. The typical speed in the NECC core is 30 cm s^{-1} at most, and the core proper tends to outcrop (from 75 to 25 m), as the current progresses to the east. The NECC axis in the eastern Tropical

Atlantic Ocean is displaced to the north due to the tilting of the ITCZ axis. East of 30°W, NECC and the Equatorial Undercurrent are separated by the west-oriented flow (the South Equatorial Current). The NECC width increases, whilst its depth decreases, as it travels farther to the east. The NECC's annual mean mass transport diminishes from 30 Sv in the area of its genesis to 20 Sv near 22°W. Farther to the east, the NECC bifurcates. The northern branch accounts for one third of the NECC's mass transport and reaches as far as 13–15°N, contributing to the formation of a quasi-stationary cyclonic feature, i.e. the Guinean dome, its center being approximately near 12°N, 22°W (Fig. 4).

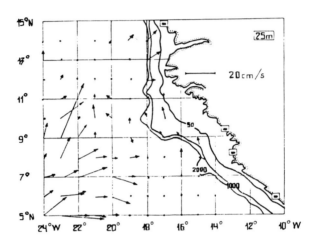

Figure 4.   Annual averages of currents based on *in situ* measurements at 25 m depth.

North of 13–15°N, this branch of the NECC barely stands out in its annual course from the background noise. The NECC's southern branch has a kinematic structure, which is similar to that of the NECC main stream in the eastern section of the Tropical Atlantic Ocean; it gives rise to the Guinean Current, which flows to the Gulf of Guinea. Circulation in the upper 300 m layer of the northeastern section of the Tropical Atlantic Ocean is twice as weak compared with the northwestern section. The NECC's mean axis in the central Tropical Atlantic Ocean nearly coincides with that of the ITCZ (Fig. 5).

The ECC's southern branch, its speed being in excess of 10 cm s^{-1}, is traceable in the subsurface ocean layer in the southern hemisphere between 4 and 8°W. The Equatorial Undercurrent takes its origin below 50 m in the vicinity of 40°W. Farther to the east, its core is observed to rise up to a depth of 25 m. The core appears asymmetrical versus the equatorial plane: the highest speeds (larger than 50 cm s^{-1}) occur south of the equator. The Equatorial Undercurrent's velocity decreases with depth, but at a depth of about 500 m south of the equator, another core of the Equatorial Undercurrent is observed, its speed being 20 cm s^{-1} (Fig. 5b). The mean annual mass transport of the Equatorial

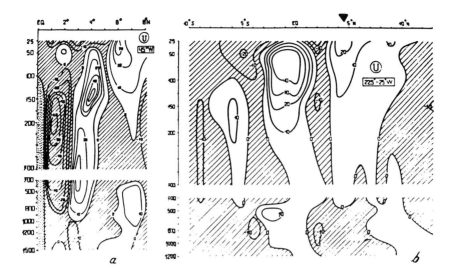

Figure 5. Annual averages of zonal currents based on *in situ* measurements at 40°W(a) and 23°W(b).

Undercurrent in the central Tropical Atlantic Ocean within a 25–1500 m layer is 45 Sv with two thirds of it being in the 300–1500 m layer. The total rate of flow of the east-oriented countercurrents in the western section of the Tropical Atlantic Ocean amounts to 90 Sv. The South Equatorial Current, which separates three branches of countercurrents, is most intensive (20 cm s^{-1}) in the top 50 m layer and slowly degrades with depth, as the compiled data indicate. The North Equatorial Current, its speed being 5–10 cm s^{-1}, travels in the upper 1000 m layer north of the NECC (Fig. 5). It seems impossible to evaluate the total mass transport of the two currents in view of the sparsity of *in situ* observations at their southern and northern boundaries.

In general, the kinematic structure of zonal currents in the central Tropical Atlantic Ocean, as provided by field observations, is similar to the one discussed in refs. [8, 11], although the total rate of flow of the east-oriented countercurrents was formerly underestimated. We will point out the absence of a meaningful quasi-stationary meridional transport of subsurface equatorial waters along the shore of South America and Africa and to the presence of quasi-stationary eddies, i.e. the anticyclonic eddy in the western boundary layer and the cyclonic one in the eastern boundary layer.

In view of the sparsity of data, we will restrict ourselves in our analysis of the seasonal variability of currents to rough description of the annual harmonic. For the central and eastern sections of the Tropical Atlantic Ocean, the semi-annual averaging will be applied. We will consider circulation in the northwestern part of the region from January to April and from June to

December, as two types of circulation are characteristic for this region. In addition, transition from one type of circulation to another one takes place during May and during December–January [7, 14]. It should also admitted that the lack of observations in the northwestern section of the region in May has also influenced the choice of an averaging interval.

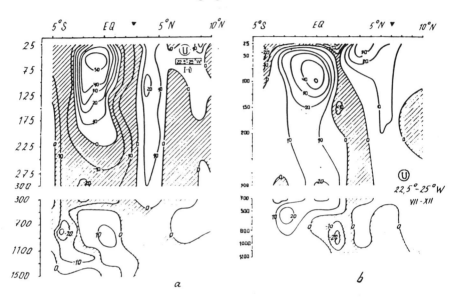

Figure 6.   Seasonal zonal currents as given by *in situ* measurements averaging between 22.5 and 25°W.

The typical velocity of the South Equatorial Current in the upper 50 m layer during the first half of the year is 10 cm s^{-1} and during the other 6 months is 20–30 cm s^{-1}. This actually confirms the intensification of the South Equatorial Current in summer indicated in ref. [7], as the majority of autonomous buoy stations were occupied during the second half of the year in this area in July and August. At larger depths, the amplitude of the South Equatorial Current's seasonal variability rapidly diminishes (Fig. 6). During the second half of the year, the North Equatorial Current travels beyond the boundaries of the considered area, although its southern border reaches 6°N during the first half-year. The typical velocity here does not exceed 10 cm s^{-1}. Seasonal variations of the equatorial countercurrents in the near-surface layer have a considerably larger amplitude, and the NECC and Equatorial Undercurrent have larger mass transport in the second half-year over the entire Equatorial Atlantic Ocean. Moreover, velocity in the main core of the Equatorial Undercurrent varies insignificantly, and an increase in the mass transport is conditioned by its modified kinematic structure; during the second half-year, the North Equatorial Undercurrent is housed by the 300–1500 m layer; during the first half-year it is divided in the 225–500 m layer by the west-oriented flow. Of all currents discussed, the near-surface NECC's intensification and widening during the second half-year are most spectacular (Fig. 6). At this time, the major contribution to the current generation in the northwestern section of the Tropical Atlantic Ocean is from the Guiana Current and the NECC's rate of flow then becomes as large as 40 Sv. Both NECC branches intensify in the eastern Tropical Atlantic Ocean and the total mass transport for the countercurrents becomes approximately twice as large during the second half-year.

Note that in contrast to the ocean surface, the NECC and the Equatorial Undercurrent are traceable in the eastern, as well as in the western Tropical Atlantic Ocean subsurface layer throughout the year. This is readily visualized, in particular, from the diagnostic computations data, which also demonstrate an essential difference in seasonal cycles of these two areas.

RESULTS OF DIAGNOSTIC COMPUTATIONS

In characterizing the intensity of zonal currents in the Tropical Atlantic Ocean based on diagnostic computations involving the quasigeostrophic model and the Princeton University density field, we will single out two principal regions*: the north region with prevailing westward transport by the Canary Current, North Equatorial Current, and the Antilles Current; and the south region where seawater is transported eastward by the NECC. Typical velocities in the north region are 5–10 cm s^{-1}, and the total transport in the 0–250 m layer attains 5–7 Sv. The largest seasonal variations of mass transport (5 Sv) in this region take place in the western and central areas. The maximum westward transport (8.7 Sv) occurs in the western part (46°W) in winter. The North Equatorial Current appears here as a fairly intensive gradient flow with the main stream being at 12°N and having a velocity attaining 15 cm s^{-1}. Toward the east, the current slackens and at 34°W, its velocity does not exceed 10 cm s^{-1} (Figs. 7 and 8). Currents and zonal transport evolving in the NECC area display large seasonal velocity variability.

The NECC is traceable in the subsurface ocean layer throughout the Tropical Atlantic Ocean region in all seasons in the form of an intensive meandering eastward flow travelling between 3 and 5°N. It is generated in the northwestern Tropical Atlantic Ocean near-equatorial region as a result of the Guiana Current's and the North Equatorial Current's turning eastward. Bifurcation of the NECC takes place in the eastern Tropical Atlantic Ocean, with the southward branch (the more intensive one) contributing to the Guiana Current, and the northward branch contributing to the formation of the Guinean dome (Fig. 7). At 46°W, the NECC is most developed during the autumn/winter

---

*Due to the geostrophic perturbations near the equator, the diagnostic circulation is considered beyond the limits of the 2° equatorial strip of water [15].

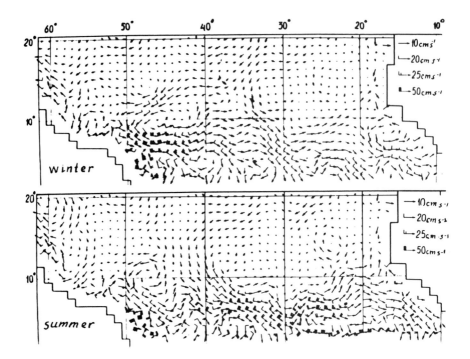

Figure 7. Currents at a depth of 50 m obtained from diagnostic calculations [15].

period. Current velocities in the NECC core at this time of the year are 50–60 cm s^{-1} and the rate of flow 25 Sv. During the spring/summer period, the NECC considerably decays. Its speed normally does not exceed 15–20 cm s^{-1}. Mass transport in spring is about 10 Sv and in summer 15–20 Sv. In the eastern Tropical Atlantic Ocean (22°W), the NECC is most developed during the summer–autumn period. The current's velocity in the core at this time of year is 20–30 cm s^{-1} and mass transport is about 10 Sv. During the winter–spring period, the current's velocity decreases to 10–15 cm s^{-1} and the mass transport to 5–7 Sv. The double-stream structure of the NECC at 45°W is particularly marked in spring and at 22°W in summer (Fig. 8).

The considered regularities of the NECC's seasonal variations are supported by the data from some cruises. It should be added that the climatic data-based (diagnostic) estimates of the NECC's intensity and mass transport are two times smaller than the actual ones. As indicated in refs. [16, 17], the NECC's mass transport varies from 10 to more than 40 Sv near 45°W and from 4 to 30 Sv in the vicinity of 30°W (Fig. 9, Part I).

A lag in meridional deviations of the NECC with respect to those of the ITCZ reaches 6 months (in the northwestern Tropical Atlantic Ocean). When

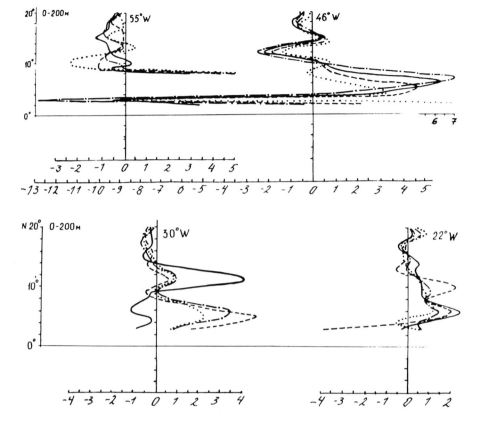

Figure 8. The currents' zonal mass transport in the upper 200 m layer [15] in winter (a) and in summer (b).

the ITCZ starts shifting south (in autumn), the NECC's mass transport keeps growing for 2 or 3 months. It then degrades moving northward and vanishes at the end of spring, when the ITCZ is located in the vicinity of the equator (Fig. 10), when another branch of the NECC emerges. Thus, it is natural to expect bifurcation of the NECC into two branches in the northwestern Tropical Atlantic Ocean subsurface layer at the end of spring.

Meridional heat and mass transport

As demonstrated above, an intensive meridional transport in the subsurface layer at the boundary separating the north tropical gyres does not take place. On the other hand, an annual mean quasi-stationary heat transfer (of the order of $1.5-10^{15}$W) in a northerly direction is known to take place in the northern Tropical Atlantic Ocean, which is provided with the vertical circulation cell [18, 19]. The mechanism responsible for this quasi-stationary heat transfer in

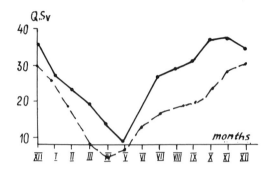

Figure 9. The NECC's mass transport as given by *in situ* measurements at 45°W [16] and 30°W [17].

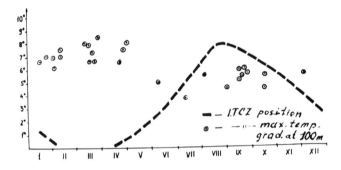

Figure 10. The ITCZ's position and the largest temperature gradient at the depth of 100 m near 45°W [4].

the northern section of the Equatorial Atlantic Ocean, as indicated in ref. [20], is the transport of warm surface waters northward by drift currents and the compensating south-oriented barotropic transport of cold deep waters. The results obtained support this inference. Integral drift meridional mass transport between 10 and 15°N varies in the course of the year from 10 to 25 Sv, which amounts to a mean annual value of about 18 Sv. Baroclinic meridional transports in subsurface layers at those latitudes are nearly zero (Fig. 11). The appropriate meridional heat transfer can be readily estimated via multiplication of the given rates of flow by the difference between temperatures in the upper mixed layer and the mean ocean temperature at these latitudes (about 20 °C). The mean annual heat transfer is $\sim 1.4 \times 10^{15}$W. Thus, the presented mechanism may provide for nearly the total heat transfer from the Equatorial Atlantic Ocean.

It will be appropriate to note an appreciable seasonal variance of integral meridional gradient mass transports between 5 and 10°N in the vicinity of the

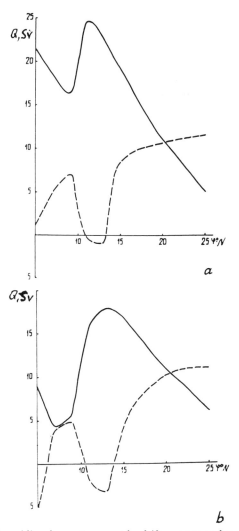

Figure 11. Integral meridional mass transport by drift currents and gradient quasigeostrophic currents in the upper 500 m layer (the dashed curve): (a) in winter; (b) in summer.

NECC, which is larger than 10 Sv, and weak seasonal variability of meridional transports north of 15°N. This is completely consistent with the regularities of the seasonal variability of Tropical Atlantic Ocean hydrophysical fields described above. North of 20°N, the contribution of subsurface gradient currents to total meridional transports is superior to that of the drift currents (Fig. 11), which is mainly accounted for by the presence of intensive subsurface meridional flows in the Subtropical Atlantic Ocean western boundary layer and by the attenuation of the wind's zonal velocity component in the central part of the subtropical gyre.

SEASONAL VARIABILITY OF TEMPERATURE AND SALINITY IN THE SUBSURFACE LAYER

The T, S-seasonal variability was evaluated from the monthly data contained in the Marine Hydrophysical Institute's oceanographic data-base. This has allowed confirmation of the existence of the afore-going regularities in the seasonal variability of Tropical Atlantic Ocean hydrophysical characteristics and to define these with better accuracy for the northern part of the region.

Horizontal subsurface temperature distribution in the western section of the Tropical Atlantic Ocean (40–46°W) allows identification of two frontal zones, which correspond to the main streams of the NECC and the South Equatorial Current. The southern frontal zone evolves in March–June near 3–6°N and then starts shifting north, being rather intensive until February. In March, horizontal temperature gradients become less accentuated although the front continues to shift northward and reaches 11–12°N in May. In June–July, this front virtually disappears.

The northern frontal zone (the North Equatorial Current) is annually situated approximately between 10 and 13°N. Seasonal temperature fluctuations there are less pronounced than those in the NECC's frontal zone. The North Equatorial Current front's maximum displacement to the north occurs in spring, when it approaches 13–15°N; in June–July the front becomes distinctly less pronounced.

In the central and eastern sections of Tropical Atlantic Ocean, horizontal temperature gradients in the thermocline of the NECC's and the North Equatorial Current's frontal zones are, on average, 1.5–2 times smaller than in the western section. The NECC front stands out above the noise level from May to January at 30°N and from July to February at 23°N, occupying the northmost position (8°N) in December. The North Equatorial Current front is observed from July to January at 30°W, and then in gradually shifts from 11 to 14°N. At 23°W, this front barely stands out above the noise level throughout the year (Fig. 12).

A frontal zone coinciding with the maximum of the South Equatorial Current's gradient component is documented in the central and eastern sections of the region near 5°S. At 30°W, this front is most pronounced from January to May, and at 23°W, from January to August, which is indicative of the intensification of the South Equatorial Current's gradient component during this period. It will be appropriately noted that, as indicated by the ship drift data (Figs. 1 and 2), the highest velocities in the South Equatorial Current's main stream are observed in this area in spring and summer, i.e. the phase shift between the current's seasonal variations in the subsurface layers in the area bounded by 20 and 23°W is equal to about 2 or 3 months. The largest amplitudes of the annual harmonic of seasonal temperature fluctuations in the thermocline (4 °C) occur in the western section of the region (35–45°W) between 6 and 8°N, i.e. at the NECC latitude. Temperature fluctuations in the North Equatorial Current are twice as small. In the eastern section, the

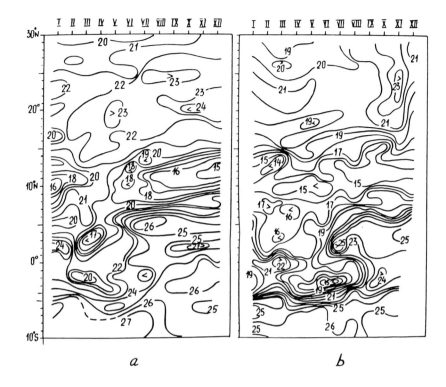

Figure 12. Seasonal temperature variability at a depth of 75 m 30° W (a) and at a depth of 50 m near 23° W (b).

largest amplitude of the annual harmonic of temperature in the NECC area is displaced closer to the equator.

The local maximum annual temperature fluctuations amplitude related to the seasonal variation of the thermal structure in the NECC zone in the eastern section is seen to be larger than 3 °C at 23°W. In the western part of the region, accommodating the NECC and the North Equatorial Current, there is almost a 6-month phase shift between the annual wind field fluctuations and the temperature field at 75–100 m depth; moreover, the phase shift diminishes as one moves eastward (Fig. 13).

Semi-annual temperature fluctuations amplitudes in the NECC zone are equal to about 2 °C in the western and central sections of the region to the east, these are less marked at 23°W, and semi-annual temperature fluctuations in the NECC's 75–100 m layer become virtually normal. The seasonal response at these depths in the North and South Equatorial Currents is most pronounced in the eastern Tropical Atlantic Ocean, where its amplitude attains 2 °C (Fig. 14).

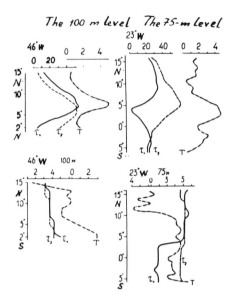

Figure 13. The amplitudes and phases of the annual harmonic of temperature fluctuations, τ_{0x} and τ_{0y}, at 46 and 23°W, respectively.

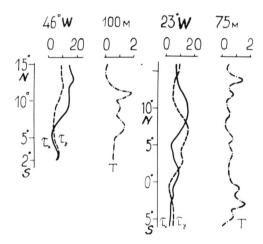

Figure 14. The amplitudes of the semiannual harmonic at 46 and 23°W.

The largest values of salinity in the subsurface layer of maximum salinity (37.5‰) are observed in the western Tropical Atlantic Ocean at the northern boundary of the studied area in July–August. During this period, the most intensive formation of brackish subtropical waters is underway in the central part of the North subtropical anticyclonic gyre.

To the south, salinity concentration drops in all seasons (up to 36–36.2 ‰ in the equatorial area); between 4 and 10°N salinity concentration rises by 0.1–0.2 ‰ during the autumn–winter period, which is related to the advection of highly-saline waters by the Equatorial Countercurrent, which intensifies at this time of the year.

Two areas accommodating large volumes of subsurface subtropical waters are identified in the western section of the north-equatorial region in spring [21]. These areas, located between 9–10° and 3–4°N, respectively, indirectly support the inference about the double-stream structure of the NECC during this period. In summer, the southern area, housing large volumes of subtropical waters, is barely pronounced and during the other seasons is not marked (Fig. 15).

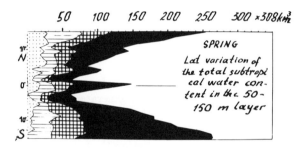

Figure 15.   The volume of subtropical waters in a 50–150 m layer ($\times 308$ km^3) [21].

The maximum concentration of subtropical waters in the equatorial region is marked in all seasons, but it is most pronounced during the spring–summer period (from March to July).

The amplitudes of annual salinity variations in the layer of maximum salinity are much smaller than in the near-surface layer and normally do not exceed 0.1–0.2 ‰. The highest salinity concentration resulting from the seasonal displacement of highly-saline subtropical waters occurs in the western area (35–45°W).

CONCLUSIONS

Complex analysis of wind, temperature and salinity, as well as the generalization of *in situ* current observations and of the calculated diagnostic data on the Tropical Atlantic Ocean currents, allow the following deductions:

(1) Seasonal variability of the hydrophysical characteristics in surface and subsurface layers of the ocean is distinctly different, this being readily seen from

considerable differences in the amplitude phase characteristics of the annual
and seasonal harmonics.

(2) The NECC's northern boundary on the sea surface replicates the Subequatorial front, and its main stream correlates with the area of maximum seawater freshening at the ocean surface. The NECC's southern boundary is best
marked in summertime by virtue of the sea surface temperature maximum, i.e.
the thermal equator located 1–2° north of it.

(3) The largest amplitudes of the seasonal variations of zonal currents and
salinity at the sea surface (30 cm s^{-1} and 2‰, respectively) and small temperature variations (about 1 °C) occur in the NECC area. The amplitude of
seasonal variations increases with depth and attains 4 °C in the thermocline,
continuing to grow larger in the westerly direction.

(4) Seasonal variability of gradient currents north of 10°N is modest within
the North Equatorial Current area, annual variations of the zonal velocity being
5–10 cm s^{-1} at most and mass transport 5–7 Sv. Seasonal variations of the
current velocity in the NECC region constitute 50–60 cm s^{-1} and the rate of
flow 20–30 Sv in the western Tropical Atlantic, and 15–20 cm s^{-1} and 5–10 Sv,
respectively, in the eastern Tropical Atlantic Ocean. The annual mean rate
of flow of the NECC, being 30 Sv at the site of the latter's origin, reduces to
20 Sv in the eastern Tropical Atlantic Ocean.

(5) The NECC's seasonal cycle in the subsurface ocean layer seems to have
the following pattern: the countercurrent is generated in the western section of
the region near the equator at the end of spring and the beginning of summer.
At this time, it forms an integral easterly flow comprising the Equatorial Undercurrent. The NECC then shifts to the north, following the ITCZ's movement, simultaneously increasing its mass transport. When the ITCZ starts
shifting in the southern direction (at the end of summer and the beginning
of autumn), the NECC's mass transport continues to grow during the course
of 2 or 3 months. It then becomes less intensive shifting to the north and
ultimately degradates at the end of spring. During the period of its decay, a
new stream of the NECC evolves in the southern area. Similarly, two NECC
streams are observed in the eastern Tropical Atlantic Ocean, but in contrast to
the western section, these are better pronounced during the summer–autumn
period, when the southwesterly trade wind intensifies, and are being controlled
by the bottom topography and coast configuration.

(6) The largest semiannual variations of temperature, salinity, and currents
at the sea surface occur in the eastern Tropical Atlantic, with their amplitudes
being 1.5 °C, 0.4‰, and 20 cm s^{-1}, respectively. The largest amplitudes of
semiannual temperature fluctuations in subsurface layers are observed in the
western section of the region, where these attain 2 °C.

(7) The subsurface oceanic layers' response to the atmospheric forcing in
the NECC area has a 3–6 month long delay; this lag tends to increase, as
the countercurrent travels from east to west, and manifests itself in both the
thermal structure and the currents' field.

(8) The total annual mean transport of water masses in the eastern direction by the system of the equatorial countercurrents amounts to 90 Sv in the western section of the Tropical Atlantic Ocean; and is observed to decrease in the eastern direction.

(9) Meridional integral drift transports between 10 and 15°N vary from 10 to 25 Sv in the course of the year, the geostrophic transports in the 0–500 m layer being nearly zero. The meridional drift transport of warm surface waters and the compensating barotropic transport of cold deep waters in the opposite direction account for the basic export of heat from the Equatorial Atlantic Ocean, which amounts to 1.5×10^{15}W.

REFERENCES

1. Artamonov, Yu. V. and Polonsky, A. B. Seasonal variability of the Tropical Atlantic Ocean hydrophysical characteristics. Part I. *Investigations of the Tropical Atlantic Ocean.* Utrecht, The Netherlands: VSP (1992), 1–19.

2. Artamonov, Yu. V. Oceanographic conditions off the northwestern African coast. *Experimental Studies of the Hydrophysical Fields.* Sevastopol: MHI (1983), 97–104.

3. Bubnov, V. A. Water circulation in the equatorial zone of the World ocean. Doctoral thesis. Moscow: (1986), 530 p.

4. Bulgakov, N. P. and Polonsky, A. B. The main hydrophysical fields of the Tropical Atlantic Ocean and their variability. Sevastopol: Preprint MHI, (1985), 44 p.

5. Zelenko, A. A., Mikhailova, E. N., Polonsky, A. B. and Shapiro, N. B. Modelling of seasonal variability of the Equatorial Atlantic currents and temperature fields. In: *The Problems of Ocean Dynamics.* Moscow: Mir (1984), 70–79.

6. Richardson, P. L. and McKee, T. K. Average seasonal variation of the Atlantic equatorial currents on historical ship drifts. *J. Phys. Oceanogr.* (1984) **14**, 1226–1238.

7. Richardson, P. L. and Walsh, D. Mapping climatological seasonal of surface currents in the Tropical Atlantic using ship drifts. *J. Geophys. Res.* (1986) **91** (C9), 10,537-10,550.

8. Khanaichenko, N. K. *The System of Equatorial Countercurrents in the Ocean.* Leningrad: Gidrometeoizdat (1974), 196 p.

9. Khlystov, N. Z. *The Structure and Dynamics of the Waters in the Tropical Atlantic.* Kiev: Nauk. dumka (1976), 164 p.

10. Belevich, R. R. Peculiarities of the variability of the Equatorial Atlantic currents. *Tr. GOIN* (1984) **175**, 34–47.

11. Bubnov, V. A. and Egorikhin, V. D. Investigations of the large-scale structure of currents in the Tropical Atlantic Ocean at 23°30′W. Moscow: *Okeonologia* (1977) (6), 19–37.

12. Artamonov, Yu. V., Bulgakov, N. P., Belous, L. M. *et al.* The main currents in the Tropical Atlantic as retrieved from *in situ* measurements. Moscow: Dep. VINITI, N 9047-B87, (1987), 28 p.

13. Polonsky, A. B. On an average circulation of the equatorial Atlantic Ocean. *TO-AN* (1987) (41), 10–12.

14. Polonsky, A. B., Goryachkin, Yu. N., Kazakov, S. I. and Pereyaslavsky, M. G. Large-scale circulation in the northwestern Tropical Atlantic and its seasonal variability. In: *Experimental studies of the Tropical Atlantic.* Moscow: Dep. VINITI, N 4986-B85, (1985), 26–37.

15. Artamonov, Yu. V. Seasonal variability of the main hydrophysical characteristics of the Tropical Atlantic. Abstract of the Doctoral dissertation. Moscow: (1989), 23 p.

16. Artamonov, Yu. V., Bulgakov, N. P. and Lvov, V. V. The kinematic and thermohaline structure of waters in the area of generation of the North Equatorial Countercurrent. Moscow: Dep. VINITI, (3743-B88), (1988), 25 p.

17. Baev, S. A. and Polonsky, A. B. Seasonal variability of the kinematic structure of waters in the central Tropical Atlantic. Moscow: *Okeonologia* (1990) (5), 83–97.

18. Hastenrath, S. Heat budget of ocean. *J. Phys. Oceanogr.* (1980) **10**, 159–170.

19. Bryan, K. Poleward heat transport by the oceans: observations and models. *Ann. Rev. Earth Planet. Sci.* (1982) **10**, 15–38.

20. Polonsky, A. B. Circulation in the Tropical Atlantic and meridional heat transfer. *Morsk. Gidrofiz. Zh.* (1985) **1**, 58–62.

21. Artamonov, Yu. V., Bulgakov, N. P. and Cheremin, V. N. Subtropical waters in the Atlantic equatorial countercurrents. Moscow: Dep. VINITI, N 9048-B87, (1987), 16 p.

Investigations of the Tropical
Atlantic Ocean, pp. 43 – 50
© VSP 1992.

The kinematic structure of the North Equatorial Countercurrent

S. A. BAEV and N. P. BULGAKOV

Abstract — The kinematic structure of the North Equatorial Countercurrent, as recovered from five large-scale hydrophysical surveys of the Tropical Atlantic Ocean during 1986–1988, is considered. Its spatial and seasonal features are identified, the characteristics of the position, depth, velocity, and mass transport are qualitatively determined.

The growing interest in the exploration of the dynamics of Tropical Atlantic Ocean waters is motivated by its important role in the mass and heat redistribution between tropical and moderate latitudes, and between the ocean and atmosphere.

As is known, circulation in the North Tropical Atlantic Ocean is controlled by the interaction of large dynamical systems in the ocean, namely, the North subtropical anticyclonic gyre, the North tropical cyclonic gyre, and the Equatorial anticyclonic gyre. Position of the main currents there depends on the location of those gyres.

As a result of studies of the tropical ocean circulation conducted in the 1960s and 1970s, the basic elements of the kinematic structure in surface and subsurface layers were identified, which are related to the North Equatorial Countercurrent (NECC), the Equatorial Undercurrent (EU), the Guiana Current (GC), the North Equatorial Current (NEC), and the South Equatorial Current (SEC) [1–3].

Investigations carried out by the Marine Hydrophysical Institute oceanographers in the northwestern Tropical Atlantic Ocean from 1981 to 1985 have stimulated a general idea about the NECC in the area of its genesis [4–6].

A series of large-scale surveys conducted during 1986–1988 have provided new data on the circulation in the North Tropical Atlantic Ocean. Analysis of the hydrophysical survey data compiled in the eastern Tropical Atlantic Ocean is presented in refs. [7, 8]. The relationship between the geostrophic circulation in the active and intermediate layers, as well as the influence of the bottom topography upon the upper ocean layer circulation is considered in ref. [9].

The present study considers the large currents in the North Tropical Atlantic Ocean and its seasonal evolution. We have derived qualitative characteristics of the NECC's kinematic structure and mass transport and have analysed the

UDK 551.465

latter's spatial variability, for which purpose the data from five CTD surveys were used. The test area is bounded by 1°N and 12°N, the eastern and western boundaries were constituted by the economic zones of the West African and South American states, respectively. The MHI-4102 probe was deployed at the joints of the 0.5° lat × 1.5° lon grid (a 0.5° lat × 3° lon grid was used in the 5th survey). Hydrophysical surveys were conducted during the following periods: the 1st survey was implemented in spring of 1986 (Cruise 33 of R/V *Akademik Vernadsky*, Cruise 46 of R/V *Mikhail Lomonosov*, and Cruise 42 of R/V *Volna*); the 2nd survey was conducted at the end of summer – the beginning of autumn of 1986 (Cruise 34 of R/V *Akademik Vernadsky* and Cruise 47 of R/V *Mikhail Lomonosov*); the 3rd survey was implemented during the winter–spring period of 1987 (Cruise 35 of R/V *Akademik Vernadsky* and Cruise 44 of R/V *Volna*), the 4th and 5th surveys were conducted during summer of 1987 and 1988 (Cruises 36 and 37 R/V *Akademik Vernadsky*, Cruises 48 and 50 of R/V *Mikhail Lomonosov* (Table 1)).

Through the use of standard techniques, we determined dynamic heights, geostrophic velocities and mass transport of currents with respect to the 800 db reference surface. The mass transports of currents were determined through the trapezoid method integration. An 0 cm s^{-1} isotach was adopted as the current's boundary. The applicability of the geostrophic relations with the purpose of defining the NECC's characteristics and the mass transport does not cause any doubt. This is corroborated by the comparison of the calculated data with *in situ* measurements conducted in the central and eastern Tropical Atlantic (the 35th and 37th Cruises of the R/V *Akademik Vernadsky*), as well as in the north western section of the region [6].

The major errors occur due to the noise of the density field rather than deviation from the geostrophic state. The choice of a reference surface in the studied region seems to be difficult because of the relatively weak horizontal circulation in the intermediate layers, as well as because of the excessive number of layers in the vertical structure. As indicated in various references [2, 10, 11], zero surface is located within a 500–2000 db range. On the other hand, *in situ* observations of currents, as well as computations of the geostrophic circulation for the 800 db surface with respect to the 1200 db one, indicate that velocities at this depth are normally 5–10 cm s^{-1} at most. Exceptions are the boundary areas, where this quantity is likely to be very large, being consistent with the level of synoptic and mesoscale noise in the region, but several times smaller than the velocity in the current's core. Besides, the estimates of the NECC's mass transport [5, 6] point out that the major water mass (80%) is transported by the countercurrent in the upper 200 m layer.

The NECC, which is the basic element of the Tropical Atlantic Ocean circulation, represents a zonally-oriented meandering flow. Figure 1 shows the maps of dynamical topography at the 50 m level, which is close by its depth to the NECC's core. The countercurrent's kinematic characteristics in its various sections are given in Table 1.

Table 1.
The NECC kinematic characteristics in the various areas of the Tropical Atlantic Ocean in 1986–1988

Surveys and dates

The NECC characteristics	Longitude °W	1 (1986) March 25–April 27 (EA) March 3–May 4 (CA) April 15–May 2 (WA)	2 (1986) July 16–Sept 16 (EA) Sept 3–Oct 26 (WA)	3 (1987) Feb 7–March 27 (WA) Feb 9–March 14 (EA)	4 (1987) June 4–July 22 (WA) June 16–August 6 (EA)	5 (1988) June 20–July 15 (EA) July 29–Sept 8 (WA)
Boundaries, °N	17.5	4–7.5[2]	7–11	2.5–5.5	5.5–9	6–10
	35.5	—	5[1]–9.5	7–10	2.5–6.5	3.5–8.5
	47.5	7–9.5	4.5[1]–9	6.3[1]–8	4.5[1]–7.5	4.5[1]–7.5
Velocity maximum, $cm\,s^{-1}$	17.5	38	33	16	50	29
	35.5	—	68	55	58	88
	47.5	29	56	58	45	75
Mass transport in a 0–200 m layer, Sv	17.5	8.8[2]	8.1	2.7	9.0	6.4
	35.5	—	15.0	8.4	19.6	19.0
	47.5	6.2	18.6[1]	7.6[1]	14.5[1]	28[1]
Mass transport in a 0–500 m layer, Sv	17.5	11.3[2]	10.5	5.0	13.2	9.7
	35.5	—	22.4	10.9	29.2	26
	47.5	7.8	24.4[1]	10.5[1]	16.2[1]	46[1]

Note: [1] The boundary of the test area;

[2] Parameters of the southern branch;

EA, WA, and CA are the abbreviations for the eastern, western, and central Atlantic, respectively.

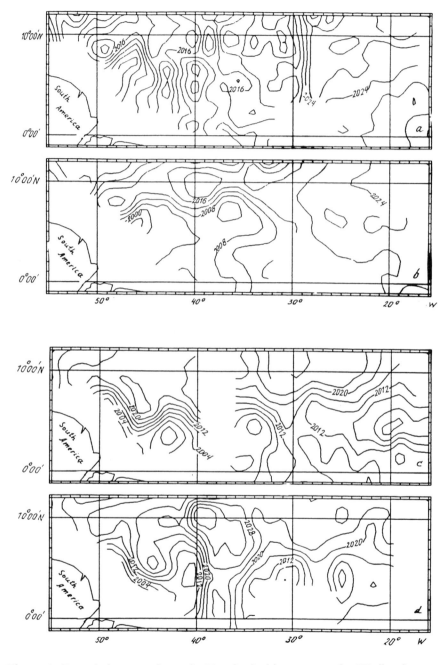

Figure 1. Dynamical topography at the 50 m level with respect to the 800 db reference surface: panels a, b, c, d, and e are constructed using the data from surveys 1, 2, 3, 4, and 5, respectively.

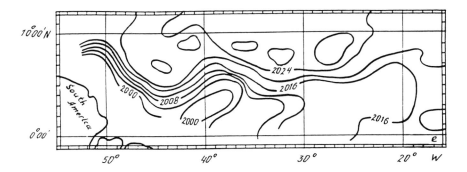

Figure 1. Continued

In spring (March–April), the NECC (survey 1, 1986) is identified as a jet flow in the western and eastern sections of the region (Fig. 1a). In the central area, only the currents of synoptic-scale variability are observed. In the area of the NECC's genesis, there exists, at this time of the year, an anticyclonic meander between 44 and 49°W produced due to the inflow of water from the north and to the reversal of the Guiana Current. The NECC occupies here the northmost position reaching 10°5′ N at 46°W. In the eastern part of the region, the surface layer current represents one flow directed at a 20° angle to the latitude.

The NECC's northern branch is observed here in the subsurface (30–250 m) layer (Fig. 2a), this being also noted by the other researchers [1, 2, 7]. Bifurcation of the countercurrent occurs at 20–27°W, with the larger water mass concentrated in the southern (main) branch.

Vertically, the countercurrent at this time of year is mainly concentrated in the top 150 m layer throughout the studied region (up to 80% of the mass transport).

The largest geostrophic velocity of 30–35 cm s^{-1} was registered at the sea surface. The NECC mass transport in spring amounted to about 10 Sv. For the northern branch, the largest velocity value is about 15 cm s^{-1} at a 40 m depth. The NECC's northern branch with a 3–3.5 Sv mass transport, possibly, represents one of the mechanisms responsible for thermal exchange between the tropics and mid-latitudes.

The 3rd survey was carried out during February–March 1987. A well-marked stream of the NECC is observable in the central part of the region at this time of the year. It occupies the northmost position here, reaching 10°N (Fig. 1b). In the area of the NECC's origin, two major eddies are singled out: an anticyclonic one, generated due to the reversal of the Guiana Current at 47–48°W and 6–8°N, and a cyclonic one, whose eastern periphery receives water of northern origin. The NECC is basically concentrated in the upper 200 m layer. In the area of its origin, as well as between 20 and 35°W, the 10 cm s^{-1} isotach

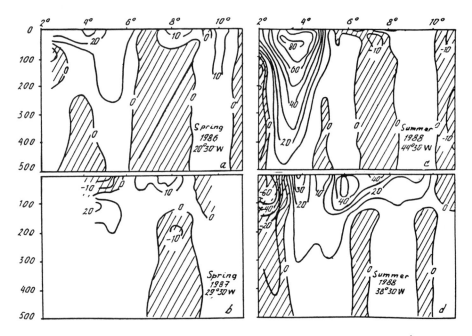

Figure 2. Vertical sections of the geostrophic velocity zonal component $(\mathrm{cm\,s^{-1}})$ in the NECC area; hatched sections indicate the west-oriented current.

goes down to the 300 m depth; and at the 90–120 m level, another velocity maximum occurs (Fig. 2b). The main countercurrent's core during this period is housed by the surface layer. The maximum geostrophic velocities and the NECC mass transport tend to decrease in transition from west to east changing from 58 to 16 $\mathrm{cm\,s^{-1}}$ and from 10.5 to 5 Sv, respectively. A specific feature of the NECC structure, as recovered from the survey data compiled in spring of 1987 and from the climatic patterns [10], consists of an appreciable decaying of the countercurrent near the region's eastern border, where the NECC did not display a well-marked stream. The computations have shown that only about 35% of the total mass transport here is eastward. The remaining mass of water contributes to the formation of the eastern peripheries of large-scale gyres observed at the countercurrent's boundaries. The NECC northern branch was not found.

The 2nd, 4th, and 5th hydrophysical surveys of the Tropical Atlantic Ocean were implemented during the summer and summer–autumn periods. Intercomparison of the derived data provides an insight into the NECC's evolution at this time of year, as there is a certain temporal shift between the surveys. On the other hand, the latter allows a rough estimate of the interannual variability of the NECC's kinematic structure.

An anticyclonic meander pitching off the NECC due to the reversal of the Guiana Current is identified in the western Tropical Atlantic Ocean (Fig. 1c, d,

e). It is gradually shifting northwestward during summer and autumn. Therefore, in June–July, it was observed at 49–50°W and 5.5–7.5°N. In September–
October, the reversal of the Guiana Current occurred at 51–53°W and 7–10°N.
The NECC's mass transport here is at its largest and is in excess of 30–40 Sv.
The countercurrent has the form of a relatively narrow (180–200 miles) and
deep stream (by the 10 cm s^{-1} isotach), which is accommodated by the upper
500 m ocean layer (Fig. 2c). The countercurrent's core is normally registered
in the subsurface layer at a 30–50 m depth. The largest geostrophic velocities
there attain 1 cm s^{-1}.

The NECC's kinematic structure in the central Tropical Atlantic Ocean is
characterized by considerable variability registered in different years. For example, during June–July 1987 it was distinctly meridionally-oriented between
37 and 38°W, and underwent a 'shearing' in the southern section of the studied
region. During August–September 1986, the NECC occupied the northmost
position (up to 9°30'N) for the summer-autumn period sinking to the 400–
500 m level. During summer 1988, the NECC was located closer to the ocean
surface, propagating in the top layer. A well-pronounced easterly transport
then encompassed the ocean region, whose width was larger than 400 miles
(from 3°20' to 10°15'N); the NECC was observed to have a double-core structure (Fig. 2d).

In the eastern part of the ocean, the NECC occupies the northernmost position during the summer–autumn period. The current is concentrated in the
top layer, the 10 cm s^{-1} isotach being observed at a 100–200 m depth. The
largest geostrophic velocity of 30–50 cm s^{-1} was documented at the sea surface.
The NECC's mass transport amounted to 10–15 Sv.

CONCLUSIONS

(1) The kinematic structure and mass transport of the NECC are characterized
by considerable intra-annual and interannual variability.

(2) In the vertical plane, the NECC normally has the form of one stream.
However, a double-core structure is likely to evolve in the central Atlantic
Ocean; and the NECC northern branch was documented in the eastern section
of the Tropical Atlantic Ocean in spring 1986.

(3) In the transition from winter to spring, the NECC is basically concentrated in the upper 200 m layer over the entire Tropical Atlantic Ocean with
the mass transport being 10–12 Sv at most. As the countercurrent becomes
less intensive during the spring period, it fails to emerge in the form of a unique
stream in the tropical zone, particularly, in its central and eastern sections.

(4) The NECC's northern branch seems to occur only in some years. Its
velocity in the core is 15–20 cm s^{-1}, and mass transport does not exceed 5 Sv.

(5) During the summer–autumn period, the NECC reaches its maximum development being one powerful stream in the Tropical Atlantic Ocean. Its width
in the western Tropical Atlantic is about 180–200 miles, and the 10 cm s^{-1} isotach's depth is 400–500 m. Here, maximum geostrophic velocities up to 1 m s^{-1}

were registered, with the NECC's mass transport being in excess of 40 Sv. In the central and eastern sections of the region, the NECC is travelling mainly in the 200 m upper layer during this period and has the maximum width of up to 400 miles. Its mass transport diminishes in the eastern Tropical Atlantic Ocean being as small as 10–15 Sv.

REFERENCES

1. Khanaichenko, N. K. *The System of Equatorial Countercurrents in the Ocean*. Leningrad: Gidrometeoizdat (1974), 158 p.
2. Khlystov, N. Z. *The Structure and Dynamics of Waters in the Tropical Atlantic*. Kiev: Nauk. Dumka (1976), 164 p.
3. Philander, S. G. H. and Düing, W. The oceanic circulation of the Tropical Atlantic and its variability as observed during GATE. *Deep-Sea Res., GATE Suppl.* (1980) **2**, 1–27.
4. Artamonov, Yu. V., Bulgakov, N. P. and Lvov, V. V. The kinematic and thermohaline structure of waters in the area of genesis of the North Equatorial Countercurrent. Moscow: Dep. VINITI, N 3743-B88 (1988), 25 p.
5. Baev, S. A., Bulgakov, N. P., Ivanov, L. I. and Kukushkin, A. S. Circulation in the area of genesis of the North Equatorial Countercurrent and its variability. *Morsk. Gidrofiz. Zh.* (1987) (4), 54–58.
6. Lvov, V. V. and Polonsky, A. B. Investigations of the structure of the North Equatorial Countercurrent in the Amazon test area. *Morsk. Gidrofiz. Zh.* (1985) (5), 58–64.
7. Baev, S. A., Bulgakov, N. P., Lvov, V. V. and Chekalin, I. A. Water circulation in the eastern Tropical Atlantic during the winter-spring period/Experimental studies of the Tropical Atlantic. Moscow: Dep. VINITI, N 90335-B87 (1987), 31–43.
8. Bulgakov, N. P. and Lomakin, P. D. Water circulation in the Tropical Atlantic during the summer season. Report on Cruise 47 of R/V *Mikhail Lomonosov* on the RAZREZY program. Moscow: Dep. VINITI, N 6828-B87 (1987), 7–19.
9. Bulgakov, N. P., Vasiliev, A. S., Efimov, V. V., Korotaev, G. K. and Lomakin, P. D. Seasonal deformation of the North Equatorial Countercurrent in the Atlantic ocean. *Morsk. Gidrofiz. Zh.* (1989) (5), 12–18.
10. Artamonov, Yu. V., Polonsky, A. B. and Pereyaslavsky, M. G. Investigations of the large-scale circulation in the northeastern Tropical Atlantic. Moscow: Dep. VINITI, N 992-B87 (1987), 25 p.
11. Stranma, L. Potential vorticity and volume transport in the eastern North Atlantic from two long CTD-sections. *Deut. Hydrog. Zeit.* (1984) **4** (37), 147–155.

Investigations of the Tropical
Atlantic Ocean, pp. 51 – 72
© VSP 1992.

The problem of variability and interrelation of atmospheric and oceanic phenomena in the Tropical Atlantic Ocean: Part I

G. S. DVORYANINOV, V. N. EREMEEV and A. V. PRUSOV

Abstract — The paper addresses the problem of the relationship between the genesis of large-scale perturbations in both the Equatorial Atlantic and the North Equatorial Countercurrent (NECC) region and the meridional deviations of the Intratropical Convergence Zone (ITCZ). It is concluded from the analysis of the air pressure field data, remotely sensed ocean surface temperature fluctuations, meridional shifting of the ITCZ, buoy-derived data on temperature and currents, as well as on the drifting buoy trajectories, and from the theoretical considerations that the period of generation of large-scale anomalous disturbances is dependent on the ITCZ's kinematics. Large-scale perturbations occur at a time when the ITCZ alternates the sign of its meridional displacement and are normally accompanied by the development of large-scale near-equatorial upwellings, sharp zonally-extended fronts, and long waves propagating westward. The latters' periods and lengths remain within $T \approx 15$–30 days and $\lambda \approx 900$–1000 km, respectively. At this time, thermohydrodynamical characteristics in the near-equatorial zone and the heat budget of the upper ocean layer drastically change.

INTRODUCTION

In recent years, the problem of air–sea interaction in the tropical oceans has been a part of all major oceanographic and meteorological programs; moreover, a number of projects were dedicated to it. In the major RAZREZY program, aimed at the study of the relationship between the air–sea interaction and climatic perturbations, investigations of the contribution of the Tropical Atlantic Ocean to weather variations in its northern regions are also considered to be of primary importance.

Such concentration of scientific effort on the study of dynamical processes in the tropical areas is related to the fact that the latter represent the central factor which controls, to a large extent, the weather variability. The tropical areas play a key role because it is there that the atmosphere receives a vast amount of thermal energy and transports it to the higher latitudes, thus compensating the long-wave radiation flux there. Moreover, the tropical oceans function as

UDK 551.460:629.78

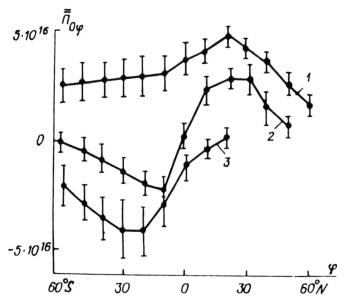

Figure 1. Meridional heat transport by currents and due to the microturbulent exchange (φ, deg) in the World's oceans recovered from the data in refs. [1–3].

energy accumulators, whereas energy consumption in polar regions, conversely, is larger than the amount of heat received from the sun. The major role of the tropical ocean (particularly, in the northern hemisphere) in the evolution of heat balance is shown in Fig. 1.

Incidentally, the largest contribution to the heat transfer from tropical regions to the northern latitudes is from the Atlantic Ocean.

An important point which determines the entire thermohydrodynamics of the tropical atmosphere is related to the fact that the Coriolis force vanishes at the equator; it changes its sign in transition across the equator, being negligibly small in the vicinity of it. However, the changing of the Coriolis force is crucial here. As a result of this, a relative maximum of westerly wind stress and its considerable meridional variability occur in the equatorial atmosphere, which conditions an occurrence of wind vorticity acting on the sea surface and inducing integral transport in the ocean. As a result of the joint effect of the above factors, powerful deep-water countercurrents and an area of elevation of deep waters, along with a number of long wave perturbations (Kelvin waves, inertial-gravity waves, etc.), evolve in the equatorial zone. All this controls, to a great extent, the interaction between the ocean and atmosphere, as well as heat transport to the north.

As the vertical fluxes and the convective processes are more pronounced in the equatorial area (likewise in the entire tropical region) than in the remaining atmosphere, the friction layer there is more powerful and, consequently, the atmospheric dynamics, the study and interpretation of the observed phenomena, prove to be much more complicated. Besides, the situation is aggravated by the presence of intensive divergence and convergence of horizontal air fluxes. It is right in the tropical region that an anomalous convergence of global air fluxes occurs, forming the Intratropical Convergence Zone (ITCZ), which affects the dynamics of the ocean and atmosphere not only in the tropical regions, but over the entire Earth as well.

Throughout the year, the ITCZ in the Atlantic Ocean is located, with the rare exception, in the northern hemisphere. In the satellite-provided imagery it manifests itself as a narrow "belt" of powerful cumulonimbus clouds encircling the Earth near the equator. This feature is readily visualized on the image (Fig. 2) received from the NOAA-7 satellite on 30 July 1974 during TROPEX.

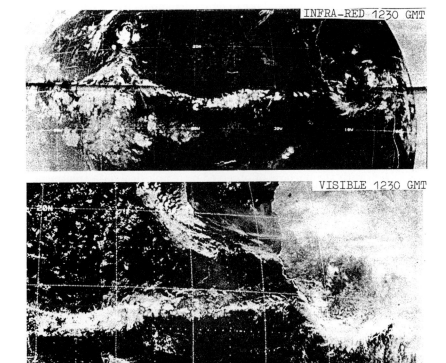

Figure 2. Satellite-derived imagery of cloudiness in the ITCZ area [9].

Aside from the large-scale phenomena mentioned above, which take place in the near-equatorial region of the ocean, and from the ITCZ in the atmosphere, a central role in the interaction of the atmosphere and Tropical Atlantic Ocean belongs to the North Equatorial Countercurrent (NECC).

In fact, the tropical atmosphere and the ocean form a unique interacting system. Therefore, position in space, intensity and temporal variability of the NECC, the Equatorial Undercurrent, and generation of various anomalous phenomena must be related with the position and the rate of the ITCZ's meridional displacement. It is obvious that the physics of the whole system is extremely complicated and major insights into it may be possible only through consecutive analysis of its individual elements.

In the present paper, we will attempt to give an exposé of some of the most relevant, in our opinion, considerations and results which have been acquired to date. A research program will then be outlined for exploration of one of the least studied phenomena of the Tropical Atlantic Ocean, involving the crucial reconstruction of the equatorial ocean dynamics and its large-scale destabilization in certain seasons.

SOME PECULIARITIES OF THE RECONSTRUCTION OF THE HYDROMETEOROLOGICAL FIELDS DEPENDING ON THE ITCZ'S KINEMATICS

Relationship between the wave field variability in the atmosphere and the ITCZ meridional displacement

A multidimensional variant of the maximum entropy method was developed in refs. [4–6] with the purpose of observed meteorological data treatment and analysis. It was used to analyse the seasonal and interannual variance of the density field spectral characteristics over a large variety of periods and of the temperature of the sea surface layer depending on meridional migrations of the ITCZ [6–10]. A number of unknown phenomena were then revealed: (i) correlation between variability of the sea surface temperature field and the ITCZ shifting with respect to the equator; (ii) existence of an interrelated quasibiannual cyclicity of both phenomena; and (iii) restructuring of the atmospheric wave field depending on the phase (season of the year) of the ITCZ meridional shifts [8–10]. The latter effect seems to be most spectacular since it occurs, as will be demonstrated below, concurrently with the dynamic destabilization in the equatorial Tropical Atlantic Ocean.

The above-indicated effects are considered in detail in refs. [6–10]. Here, we will submit only the results which are directly related to the considered problem. We will focus mainly on the study of mesoscale wave processes, their temporal evolution and restructuring depending on the ITCZ's meridional displacement. Besides, the relationship between sea surface temperature fluctuations and meridional shifting of the ITCZ axis is being analysed.

Investigations of nonzonal synoptic-scale wave perturbations, along with the elucidation of physical mechanisms controlling the intensity of these processes constitute an important objective in the study of air–sea interaction. At the same time, it should be recognized that these type of waves virtually have not been investigated. Multiple publications, including fundamental studies [11,

12], dealt solely with zonal waves. However, analysis of the data and of the methods of their processing, which led to the deduction that disturbances in the tropical areas are propagating zonally, makes one sceptic as to the validity of this inference. Firstly, a close correlation between meteorological elements in tropical areas and those in the middle latitudes was pointed out in some papers [13, 14], a fact which implies an existence of the meridional interrelation and the transport of perturbations. Secondly, in the case of nonzonal synoptic wave propagation, identification of the wave vector meridional component is complicated due to considerable spatial inhomogeneities of the meteorological fields. In order to reduce their impact on space–time spectral estimates, it is necessary to diminish meridional dimensions of the region (the antenna array), within which the measurements are being conducted. Finally, a regular alternation of the wave propagation direction may cause an illusion as to the absence of the wave field meridional components variability, if estimation involves the records of the length of $N/T \geqq 1$, where T is the characteristic period for direction reversal. However, the treatment of short records using conventional techniques of the Blackman–Tjuky type leads to an essential growth of confidence intervals thus making spectral estimates invalid, and the method suggested in refs. [4–6] permits analysis of synoptic-scale wave perturbations using short records recovered from the small-aperture antenna array.

As satellite [15] and *in situ* observations [7, 16] indicate, the area between 5–15°N and 5–20°W displays an enhanced wave activity on synoptic scales. It is well pronounced on satellite–derived pictures of cloudiness (see Fig. 2).

Figure 3. The area of active meridional transport. Isolines indicate the spectral density of the meridional wind speed.

The ITCZ is also readily visualized there. Figure 3 taken from ref. [17] shows isolines of spectral density maximums of the wind speed meridional component at 650 mb with the period range being 3.1–4.0 days, which were constructed using the data collected from July to September 1979. The center of this energetically-active zone is located at 13°N, 15°W. This zone encompasses the territory of the Republic of Guinea, whose weather stations provided the relevant data. These have the form of synchronous biannual series of the near-surface air pressure field (from 1978 to 1979), sampled every 3 h by five weather stations at Conakry (9.5°N, 13.7°W), Boké (10.9°N, 14.3°W), Mamoú (10.4°N, 12.1°W), Cancan (10.4°N, 9.3°W), and Nzérékoré (7.7°N, 8.8°W). The data were being prepared for treatment with the proviso borne in mind that the whole of the studied period would be divided into subperiods: 2–20 and 20–365 days, respectively. To eliminate the effects of processes with cycles ranging from 20 to 365 days on the first period, the high-frequency filtration through the cosine filter ($P = 104$) was performed. The cosine filter's weight coefficients were calculated as follows

$$V(K) = \frac{1}{2P}\left(1 + \cos\frac{\pi K}{P}\right), \qquad |K| \leqq P.$$

Besides, a low-frequency filtration was conducted with the purpose of eliminating highly-energetic pressure field components with cycles of 1 and 0.5 days. For this, a cosine filter ($P = 8$) was applied, which has the lines of absolute suppression at frequencies of 1 and 0.5 days, with the sampling interval being 3 h. The filtered series were subjected to rarefication, which was not expected to affect the data quality in the studied frequency range and permitted to essentially save computational time. In the ultimate form, the series had a 1400-point length with a 0.5-day discreteness (the Neiquist frequency was $F = 1 \text{ day}^{-1}$). According to the pattern of data treatment, sections of 1-month duration were subsequently handled.

The data processing and subsequent analysis revealed an instability of individual wave component within this frequency range. These proved to change and periodically evolve and disappear depending on the time of the year. The initial analysis of the estimates of $\rho(f, k)$ implied the tracing of the temporal evolution of individual wave components from the time of their emergence until their complete disappearance. In some cases, the waves' lifetime was as long as 2 or 3 months. However, in the majority of cases, wave parameter variations turned out to be sizable in transition from one month to another. It may be supposed that this effect is conditioned by the spatially-inhomogeneous and nonstationary-in-time wind field, which is responsible for nonuniform Doppler shift. Therefore, a stage involving computations and analysis of autospectra, coherences, and phase shifts between the antenna array elements was also of primary importance, as it permitted control of uncertain cases when peak interpretation in the wave number domain was ambiguous. Such analysis was imperative in view of the following. Firstly, the antenna array characteristic failed to allow a detailed study of incident wave energy angular distribution. Therefore, this called for a special approach to their analysis, when emphatically pronounced spatial inhomogeneity was involved, which manifested itself in the notable diversity of wave amplitudes of the same frequency in different elements of the antenna array. Secondly, we had to sort out cases with spatially-inhomogeneous Doppler shift. Analysis of the autospectra variability is given in refs. [7, 18].

For better control of the cases with the spatially inhomogeneous Doppler shift, one should bear in mind that the absence of an explicit frequential shift between autospectra peaks corresponding to different antenna array elements is not a definite indicator of the absence of Doppler shift. In the course of concurrent processing the data from several meters, the peaks' position in one autospectrum may affect the position of the adjacent peak in the other autospectrum. Naturally, the more energetic peak is expected to be predominant. Therefore, coherence seems to be the best criterion, as it lowers in the case of a frequential shift camouflaged by the peaks' mutual attraction.

Figure 4 shows the results of evaluating the autospectra, $S(f)$, coherence, $\varkappa(f)$, and four frequential slices of the space–time spectra (July 1978). This month was chosen in order to emphasize the possible discrepancies in autospectra and coherences, which occur at smaller periods. It is demonstrated in Fig. 4 how the averaged spectral estimates, $\overline{S}(f)$, seasonally modifying over a period of 2 years. These estimates were derived using the relation

$$\overline{S}(f) = \frac{1}{5} \sum_{m=1}^{5} \widehat{S}_{y_m}(f)(\widehat{\sigma}_{y_m}^2)^{-1},$$

where $\widehat{S}_{y_m}(f)$, $\widehat{\sigma}_{y_m}^2$ are spectral estimates and sampled dispersions calculated from the simulations of $y_m(n)$ for each observational point. It is seen that there exists a seasonally-recurrent evolution of spectra, which undergo a substantial deformation in transition from winter to summer. The well-marked peaks pertaining to the winter–spring period ($f = 0.08, 0.13, 0.23$ day^{-1}) become smoothed by the summer (ITCZ occupies its northernmost position) and the total wave energy decreases. In a year, the shape of the spectrum, with the exception of minor details, restores itself.

In view of the above, in studying the wave field's spatial restructuring dictated by the ITCZ migration, the waves, whose coherence at least at one of the grid bases was less than 0.99, were not being considered in order to avoid noise corruption of the restructuring procedure by the nonuniform Doppler shift. The wave estimation results "purified" in this manner are presented in Fig. 5a. This method does not permit tracing of the evolution of individual components; however, it depicts the principal peculiarities of the seasonal wave variability.

To control the presence or absence of angular energy distribution at the chosen frequencies, comparison of the estimates of $\rho(f,k)$ using normalized and non-normalized spectral matrices was conducted. The normalized spectral matrix is supposed to read

$$S_n(f) = \Gamma^{-1}(f)S(f)\Gamma^{-1}(f),$$

where $S(f)$ is the original estimate of the spectral matrix and $\Gamma(f)$ has the form

$$\Gamma(f) = \text{diag}\left\{S_{11}(f), \cdots, S_{nn}(f)\right\}.$$

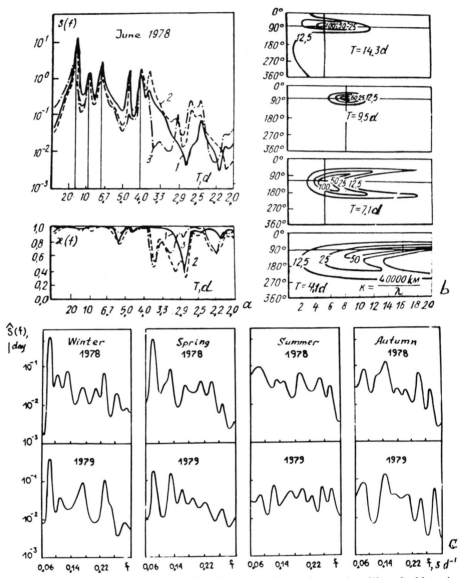

Figure 4. Autospectra estimates, $S(f)$: (1) at the Boké weather station; (2) at the Mamoú weather station; and (3) at the Nzérécoré weather station. Estimates of the coherence spectra, $\varkappa(f)$; of the sections of space time spectra (July 1978); and of temporal variability of mean spectra $\widehat{S}(f)$.

Obviously, matrix $S(f)$ can be presented in the form $S_n(f) = (S_{\alpha\beta}^{(0)}(f))$, where $S_{\alpha\beta}^{(0)}(f) = \varkappa_{\alpha\beta}(f)\exp\{i\varphi_{\alpha\beta}(f)\}$, $\varkappa_{\alpha\beta}(f)$ and $\varphi_{\alpha\beta}(f)$ are the spectra of coherence and of phase shift between elements α and β of the antenna array.

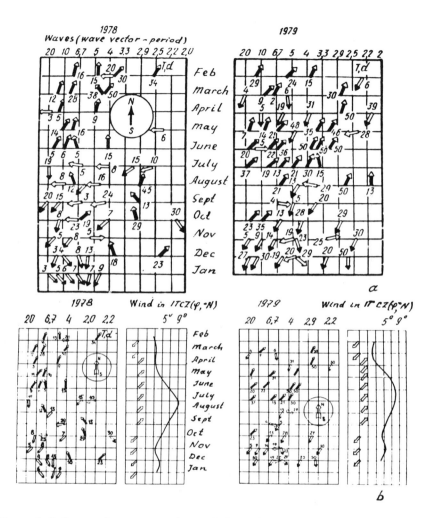

Figure 5. Estimates of vector waves (arrows indicate the direction and numerals indicate the moduli of the global wavenumber) $|k| = 40000/\lambda$ km: (a) cases with wide angular distribution; and (b) cases without wide angular distribution.

Transition to the estimate $\rho_n(f,k)$ constructed through the replacement of matrix $S(f)$ by $S_n(f)$ leads to the concentration of energy from the entire interference set at some 'mean' direction with some 'mean' wavelength. The algorithm and appropriate analysis are discussed in detail in refs. [4, 8].

If the spectrum $\rho(f,k)$ incorporates only one wave at the given frequency and its width depends only on the noise level, then estimates of $\rho(f,k)$ and $\rho_n(f,k)$ differ insignificantly. Large differences in the structure of these estimates point out to the existence of an interference at the given frequency even when a straightforward separation of individual components in the interference

set seems impossible. With this consideration , we eliminated from the data
in Fig. 5a all wave components which represent interference sets (Fig. 5) and
whose separation by the angle and modulus of the wavenumber was not possible. The ultimately 'purified' results of wave parameter estimation are shown
in Fig. 5b.

The main effect displayed is the existence of steady wave groups with large
meridional components occurring at those periods when the ITCZ's axis is
shifting in the meridional direction. The general direction of wave propagation
in the groups coincides, in general, with that of the ITCZ axis. For comparison,
the diagrams of the ITCZ axis's position in the annual cycle over 1978–1979,
based on satellite-provided cloudiness data, are plotted in Fig. 5b.

As is seen, from February to May 1978 and from March to June 1979, i.e.
when the ITCZ was shifting from south to north, there was a regular wave
field in the atmosphere which moved in the same direction, i.e. to the north.
When the ITCZ reaches its northernmost position and alternates its sense of
movement, destabilization of the wave field occurs and it starts restructuring.
The number of meridionally-oriented wave components drastically reduces and
zonal waves are generated. Waves of different orientation are observed at this
time. Then, as the ITCZ is travelling southward, the wave field undergoes
another restructuring. The majority of waves propagate from north to south.
Finally, from November to February, when the ITCZ approaches the equator
and alternates its sign of meridional displacement, wave field destabilization
takes place anew, and the waves' movement becomes essentially chaotic. However, restructuring of the wave field during this period is of longer duration.
In addition the ITCZ remains for a longer period in the southern hemisphere,
its speed of movement being smaller on average, than in the vicinity of its
northernmost location.

It can be concluded that we must differentiate the seasons during which meridionally-oriented and chaotic waves occur. Moreover, these two states of the
wave field are coupled with the kinematics of the ITCZ; so the predominantly
meridional direction of wave propagation takes place when ITCZ's meridional
shifting is permanently northward or southward. The wave field destabilization and its subsequent restructuring occur at those moments of time when
the ITCZ is near its meridionally extreme position and alternates its sign of
meridional displacement.

RELATIONSHIP BETWEEN THE TEMPORAL VARIABILITY
OF THE OCEAN TEMPERATURE AND THE ITCZ'S
MERIDIONAL MIGRATIONS

In this section we will discuss the results of analysis of mutual spectral characteristics, seasonal and interannual evolution of the ITCZ meridional shifts,
and oceanic temperature data [6, 9, 20].

For this purpose, we used METEOSAT data [19] on the ITCZ's monthly
mean meridional position ($\varphi°$) at 28°W and on the monthly mean ocean surface

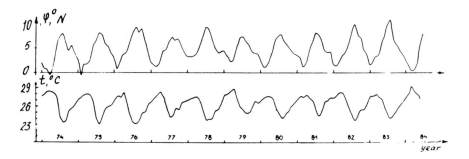

Figure 6. Relationship between the ocean surface temperature ($t°$) variability and the ITCZ's meridional displacements ($\varphi°$).

temperature ($t°$) data averaged over a meridional section from 2°N at 10°W in the equatorial Tropical Atlantic Ocean (1974–1984) (Fig. 6). The study of these characteristics on synoptic variability scales (from a few hours to several days), was conducted using TROPEX-74 data.

Initially, the annual course averaged over an 11-year period was removed from the $t°(t)$-and $\varphi°(t)$-series. Deviation from this annual course is shown in Fig. 7a.

Two inferences can be made from the analysis of these diagrams. Firstly, there exists a regular interannual variability in the behaviour of anomalies in the ITCZ meridional shifting and ocean temperature anomalies. Secondly, the long-term (several years) variability of short-term (several months) anomalies of $t°$ and $\varphi°$ occurs with the phase shift close to π, i.e. virtually in a counterphase. This is readily visualized from the figures, some of which have darkened sections.

To distinctly identify the interannual anomaly modulations, the annual means of $t°$ and $\varphi°$ were calculated with a subsequent shifting of the 1-month long averaging interval over the entire observational time. The acquired result is presented in Fig. 7b. It shows that anomalies are subjected to the strongest modulations with biannual cycle. The quasibiannual cyclicity of temperature anomalies variation occurs virtually in counterphase with the variability of the ITCZ's meridional migration. However, a failure in phase shifts was registered during 1974–1975 and 1979–1980. Naturally, this effect can be accounted for by the existence of quasibiannual cyclicity in atmospheric circulation and the failure of its phase. As is shown in ref. [21], this assumption is supported by the irregular sign alternation of the phase in the distinctly marked biannual period of oscillations of differences, δH_{2-1}, in the geopotential heights of isobaric surfaces of two subsequent years (for 10–50 mb), which took place during 1974–1975 and 1979–1980.

It also follows from the data submitted in Fig. 7a and b that in certain years the anomalies may be essentially negative, or positive. Accordingly, the

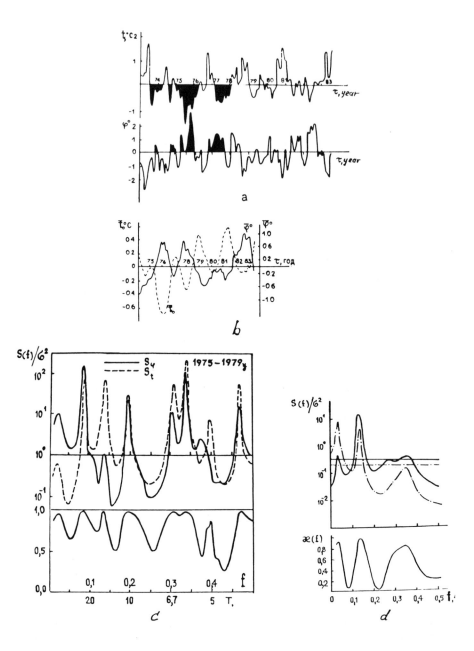

Figure 7. Deviations of Δt° and $\Delta\varphi^\circ$ from the mean annual course over a multiyear period (b); autospectra and coherence of Δt° and $\Delta\varphi^\circ$ (c); autospectra and coherence of Δt° and $\Delta\varphi^\circ$ on synoptic scales (d).

ITCZ was observed in certain years to shift northward by more than 2°, in comparison with its multiannual mean position, and there are years, as well, when the largest displacement is lesser than the annual mean.

In order to identify periods with the largest temporal variability and the frequential intervals which display meaningful correlations between $t°$ and $\varphi°$, the autospectra and the coherence spectra of those characteristics were evaluated. The results are given in Fig. 7c.

The upper diagram shows estimates of the frequential spectrum of the ITCZ displacements, $S_\varphi(f)$, the middle one portrays the ocean surface temperature spectrum, $S_t(f)$, and the lower diagram shows the coherence spectrum $\varkappa_{\varphi t}(f)$ for these parameters. The spectra were based on records, from which the mean annual course for an 11-year period was filtered out. A specific feature of the processes consists of a coinciding of the frequencies of their best marked spectral peaks for both the interannual variability and the seasonal one. This implies that within the given time intervals, ocean surface temperature variability in the tropical area is related to the meridional migrations of the ITCZ.

The largest evolution period for $t°$ and $\varphi°$, determined from the available records, amounts to 4.8 years. The next energetic peak of the highest frequency corresponds to a period of 2.2 years. Incidentally, in the first case, the phase shift between $t°$ and $\varphi°$ is 172°, and in the second case 152°.

Spectral peaks also occur in the annual periods. In the case of the annual cycle, the spectral peak had remained as the mean annual variability over an 11-year long period was filtrated, whereas the actual behaviour of $t°$ and $\varphi°$ underwent essential changes from year to year. The latter point is confirmed by the presence of spectral peaks in the 2.2 and 4.8-year long periods.

Aside from the interrelated variability of the anomalies and the coincidence of spectral peak frequencies, another argument in favour of the close dependence between the ITCZ's meridional migration and the sea surface temperature variability is the high coherence, $\varkappa_{\varphi t}(f)$, between $\varphi°$ and $t°$ at those frequencies.

To define the possible relationship between the ITCZ shifts and the sea surface temperature fluctuations on synoptic scales (from a few hours to several days), a similar spectral processing of data on the ITCZ's southern border meridional displacements at 25°W (the satellite-derived data) and daily means of the ocean temperature at a depth of 75 m (the area of largest vertical temperature gradients), observed during the second stage of TROPEX-74 in the area located at 5°N, 23°30'W, was implemented. These parameters were measured simultaneously [22, 23]. Autospectra of the ITCZ's meridional shifts and temperature, as well as the coherence function, are presented in Fig. 7d. It is seen that the above processes are closely interrelated on synoptic scales as well. This fact is particularly evident at 28- and 7.5-day long periods.

Thus, spectral analysis of the synchronously-derived satellite information on the ITCZ's meridional position, depending on the time and variability of the Tropical Atlantic Ocean surface temperature has shown that on a wide temperature scale ranging from several days to 5 years, ocean temperature fluctuations depend on the ITCZ's meridional position.

SEASONAL PERTURBATIONS IN THE TROPICAL ATLANTIC OCEAN AND THEIR RELATIONSHIP WITH THE ITCZ'S KINEMATICS

As the ocean dynamics and variability mainly depend on the tangential wind stress, air pressure, and heat fluxes through the ocean surface, the atmospheric effects described above are expected to affect the stability of oceanic phenomena. In fact, when the ITCZ is shifting northward, the line of trade wind convergence (i.e. peculiarities in the distribution of the wind stress, the air pressure field, and thermal fluxes related to the ITCZ) is also moving in a northerly direction. With respect to the ocean dynamics, these characteristics represent an external effect. Thus, with the ITCZ shifting northward, spatial distribution of the external forces upon the equatorial ocean continually varies in time, which is supposed to initiate a consecutive permanent adaptation of the ocean to this variable forcing. When the ITCZ reaches its northernmost position and starts moving in the reverse direction, the sign of the atmospheric forcing direction also alternates. As a result, the air dynamics are supposed to restructure in such a fashion that would provide for its internal balance and the latter's variability to be in accord with the ITCZ's movement to south. Thus, it is at this time that the equatorial Tropical Atlantic Ocean is expected to generate major disturbances and instability. However, since the equatorial region accommodates the inertial boundary layer, where even minor perturbations produce a relatively strong response, it is in its vicinity that the most pronounced destabilization of the 'stabilized' regime takes place. Powerful upwellings, large waves, the Equatorial Undercurrent variability, etc. will occur there. Notable manifestations of the perturbations are also expected to take place in frontal areas and jet currents, particularly in the North Equatorial Countercurrent.

Similar phenomena will also occur when the ITCZ reaches the sub-equatorial area. However, as the ITCZ is essentially less developed at this time and the rate of direction changing here is normally smaller than in the north, destabilization effects in the ocean in this case must be less pronounced.

Obviously, the definite response as to what extent the variability of natural processes controls the mechanism for air–sea interaction in the tropical ocean warrants a long-term *ad hoc* program aimed at complex monitoring of the ocean's variability and at the concurrent registering of the ITCZ's kinematics and dynamics. Currently, implementation of such a venture seems hardly possible due to the lack of necessary hardware, but the problem of investigating the Tropical Atlantic Ocean dynamics, as was indicated in the first section, is of the utmost importance and the realistic *modus operandi* here seems to be a consecutive study of the system's chief elements with a sequel-up generalization of the derived data.

As the climatic variability is strongly dependent on anomalous processes taking place in the ocean, we consider that exploration of anomalous, destabilizing phenomena currently constitutes the basic physical problem of air–sea interaction in the Tropical Atlantic Ocean. In connection with this, we will submit a program in Part II, whose implementation was started in January 1987 during Cruise 35 of R/V *Akademik Vernadsky*, and theretofore we will discuss new relevant data compiled during Cruise 50 of R/V *Mikhail Lomonosov*.

First, we will demonstrate, using historical data, that there exist multiple data which support the inferences presented above. In the course of implementation of FOCAL/SEQUAL programs dedicated to the exploration of the tropical ocean and its seasonal variability, a number of important results have been obtained relating to the problem at issue [24–28]. Genesis of a large-scale instability in the eastern Equatorial Atlantic Ocean in June–July 1983, which gave rise to the generation of a powerful equatorial upwelling, energetic long waves propagating westward along the equator and sharp meridional gradients of the sea surface temperature is discussed in refs. [25–28]. The area of instability was rapidly travelling to the west. The genesis and evolution of this large-scale instability were being monitored by satellite. Two satellite-provided images are shown in Fig. 8.

Here, the waves are readily visualized at the upwelling's northern front [26]. Observations of wave disturbance made possible for Legeckis and Reverdin [29] to estimate the length of waves, their magnitude, and direction of their phase velocity. It turned out that the waves' period was about 24 days, the length $\sim$ 1000 km, and phase velocity varied from 42 km d^{-1} in June to 18 km d^{-1} in August.

In the area of instability, the upwelling and waves travelled approximately as far as 25°W, being most pronounced during the summer period (in June–July) and then gradually vanishing. A similar crucial restructuring of the dynamics during the same period with more pronounced long waves was observed in the Pacific Ocean. The satellite data were analysed in conjunction with the drifting buoy data; it should be added that the obtained results proved identical [29]. The information given in ref. [27] also supports the argument that large-scale energetic perturbations that affect the overall dynamics of the Tropical Atlantic Ocean evolve at this time of the year. That paper contains the charts of ocean surface temperature based on the experimental data, which demonstrate the latter's rapid evolution in the equatorial ocean region during the disturbance genesis, followed by upwelling generation. The maps presented in Fig. 9 show that the disturbance started to develop in the second half of May 1983 and in the second half of June 1984, i.e. exactly at the time when the ITCZ reached its northernmost position.

Figure 6 shows that at the end of May 1983 the ITCZ shifted from the equator northward by more than 8°. In that year the ITCZ's displacement was anomalously large (in excess of 11°) and, obviously, this accounts for the early occurrence of a cold disturbance in the Equatorial Atlantic Ocean more powerful than in the year of 1984. Note that remotely-sensed data, as well as the *in situ* observations, show that strong meridional ocean surface temperature gradients evolve in the course of perturbation development.

Furthermore the data on the seasonal variability of currents' temperature and velocity near the equator at 28°W are presented in ref. [28]. Observations were conducted continually in the upper ocean layer for a period of 2.7 years.

G. S. *Dvoryaninov* et al.

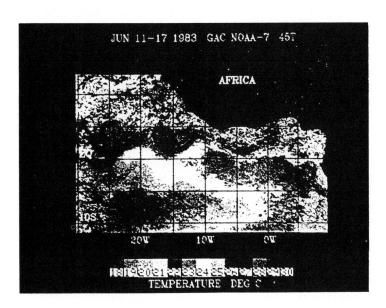

Figure 8. Manifestations of equatorially-unstable waves in the ocean surface temperature field. Satellite-derived imagery from ref. [26].

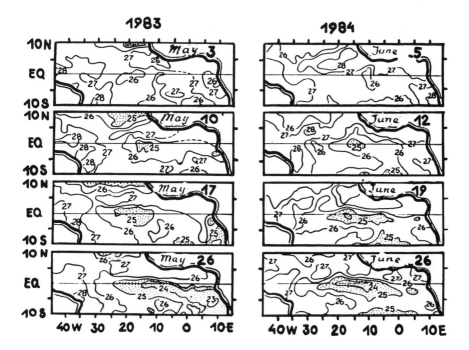

Figure 9. Charts of sea surface temperature distribution during the upwelling generation period in 1983 and 1984.

Figures 10 and 11 portray some results derived from this work. Temporal variability of isotherm depths is shown in Fig. 10a. These diagrams may also be interpreted as temperature variability with time at fixed depths.

The isotherms' behaviour indicates that annually deep waters in the equatorial region are outcropping from May to July concurrently reducing the heat budget of the upper ocean layer. The heat budget temporal evolution is depicted in Fig. 10b. The latter displays the well-pronounced annual cycle of the heat budget with the smallest values, resultant from the generation of instability and upwelling from May to June.

Distribution of the zonal velocity component over depth is shown in Fig. 11a. The solid curves are the 80 cm s^{-1}-isotachs of the east-oriented current velocity component, i.e. these represent contours of the Equatorial Undercurrent.

Analysis of the isotachs" behaviour in time shows that during the season of development of intensive upwelling and long waves, the Equatorial Undercurrent shifting in the vertical passes its mean depth and continues to go deeper. Simultaneously, the local minima of the integral-over-depth zonal (Fig. 11b) transport, and the meridional one with regular sign-alternating, large-amplitude oscillations (Fig. 11c) take place. These are caused by the generation of unstable waves. From December to January when the Equatorial Undercurrent shifting upward crosses anew its mean depth, a relative

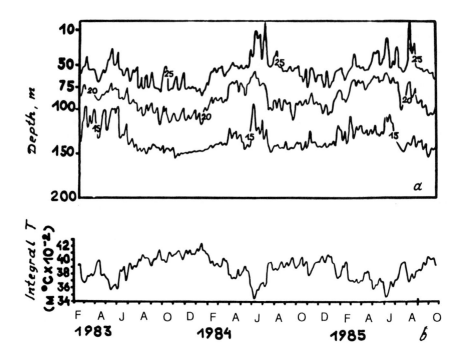

Figure 10. Temporal variance of the depth of equatorial isotherms (a) integrated over the depth of temperature (heat budget) (b).

minimum of the zonal transport and a maximum oscillation of the meridional transport occurs again, the ITCZ being located at this time near the area of its minimum meridional displacement. As is seen from Fig. 11c, the occurrence of waves manifests itself not only through the relative maxima of the meridional transport variation and through the minima of the zonal transport, but as well through the frequential modulation of oscillations. It is readily visualized from the meridional transport diagram that during the period of unstable wave generation, more pronounced is the low-frequency variability with the 1-month cycle, whilst in the other seasons it has a higher frequency.

    Drastic dynamical restructuring, accompanied by the generation of long zonal waves, occurs not only in the equatorial region; similar perturbations are also simultaneously generated (or, more exactly, are likely to be generated) in the North Equatorial Countercurrent area. On the satellite-derived colour picture (Fig. 8) one can see, concurrent with the wave perturbation in the equatorial area, a marked waves disturbance near 5°N with relatively large meridional sea surface temperature gradients and the same wavelength.

    An occurrence of such oscillations in the NECC area during this time of the year is also demonstrated in ref. [30]. The data on the variability of temperature, meridional (v) and zonal (u) current velocity components, as well as the

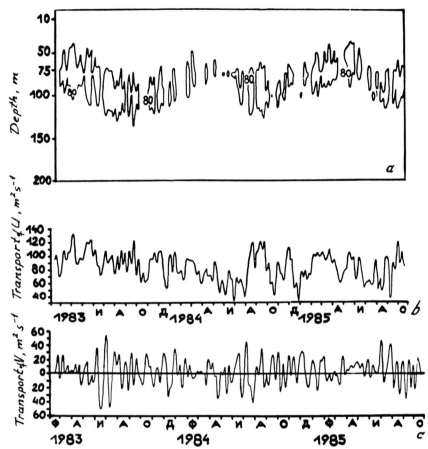

Figure 11. Temporal variance of the isotach depth of the east-oriented wave velocity component near the equator (a), of the zonal (b), and meridional (c) transport integrated over depth.

trajectory of the drifting buoy, which provided this information are presented in Fig. 12. Here, the NECC's destabilization during June–July of 1983 is clearly seen, as well as the buoy's direction changing from westerly to easterly and the respective transformation of its smooth trajectory into a wavelike curve. According to our estimation, the Lagrangian period and wavelength equal 15 days and 700 km, respectively. Finally, it will be noted that an occurrence of similar wave disturbances has been proved theoretically in ref. [31]. Figure 12b taken from this study demonstrates that, theoretically, long waves in the model seasonal cycle of the Tropical Atlantic Ocean dynamics occur during the autumn period. In addition, they are approximately 1000 km long, and the seasonal cycle is 3–4 weeks. These waves are essentially inhomogeneous in space and nonstationary in time, and receive their energy from mean currents.

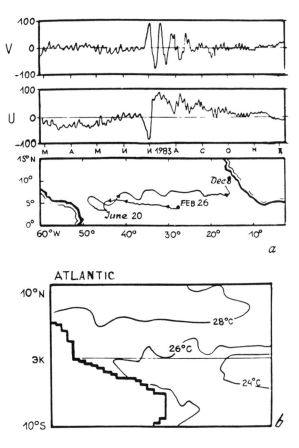

Figure 12. Manifestation of the NECC instability in the variability of the buoy's speeds, u and v, and trajectories (a); theoretically-derived vectors of the surface velocity and isotherm (b).

REFERENCES

1. Timofeev, N. A. (Ed.) *Atlas of the Oceans' Heat Balance.* Sevastopol: MHI (1970), 88 p.

2. Oort, A. H. and Vonder, T. H. On the observed annual cycle in the ocean-atmosphere heat balance over the Northern hemisphere. *J. Phys. Oceanogr.* (1976) **6**, 781–800.

3. Hastenrath, S. On meridional heat transport in the World ocean. *J. Phys. Oceanogr.* (1982) **12**, 922–927.

4. Dvoryaninov, G. S., Zhuravlev, V. N. and Prusov, A. V. The method of maximum entropy in multidimensional spectral analysis: theory and testing. Sevastopol: MHI (1987), preprint, part I, 43 p.

5. Dvoryaninov, G. S., Zhuravlev, V. N. and Prusov, A. V. The method of maximum entropy in multidimensional spectral analysis of time series. *Morsk. Gidrofiz. Zh.* (1987) (3), 3–17.

6. Dvoryaninov, G. S., Kaba, M. L. and Timofeev, N. A. The Tropical Atlantic. The Guinean region. Part I. Atmospheric processes. In: *Air-Sea Interaction*. Kiev: Nauk. dumka (1988), 7–126.

7. Dvoryaninov, G. S. The wave structure and migrations of the Intratropical Convergence Zone in the Atlantic Ocean. Sevastopol: MHI (1985), preprint, 42 p.

8. Dvoryaninov, G. S., Zhuravlev, V. N. and Prusov, A. V. The method of maximum entropy in multidimensional spectral analysis: spectral estimation of the meteorological parameters in the ITCZ area. Sevastopol: MHI (1987), preprint, part II, 22 p.

9. Dvoryaninov, G. S. and Eremeev, V. N. Some regularities in the variability of Tropical Atlantic hydrometeorological fields. *Morsk. Gidrofiz. Zh.* (1988) (3), 21–33.

10. Dvoryaninov, G. S., Zhuravlev, V. N. and Prusov, A. V. Synoptic waves in the Tropical Atlantic atmosphere and their relationship with the dynamics of the ITCZ. *Sov. J. Phys. Oceanogr.* (1989) **1**, 3–11.

11. Falkovich, A. I. *Dynamics and Energetics of the Intratropical Convergence Zone*. Leningrad: Gidrometeoizdat (1979), 246 p.

12. Dobrychman, E. M. *Dynamics of the Equatorial Atmosphere*. Leningrad: Gidrometeoizdat (1980), 288 p.

13. Mak, M. K. Laterally-driven stochastic motions in the tropics. *J. Atmos. Sci.* (1969) **26**, 41–63.

14. Julian, P. R. Some aspects of variance spectra of synoptic-scale tropospheric wind components in middle latitudes and in the tropics. *Month. Weath. Rev.* (1971) **99**, 954–965.

15. Global atmospheric research program report. (1975) (17), 180 p.

16. Burpee, R. W. Characteristic of North African easterly waves during the summer of 1988 and 1989. *J. Atmos. Sci.* (1974) **31**, 1556–1570.

17. Nitta, T. and Takayaba, Y. Global analysis of the lower tropospheric disturbances in the tropics during the northern summer of the year. Part II. Regional characteristics of the disturbances. *Pure Appl. Geophys.* (1985) **123**, 272–292.

18. Goloborodko, S. V., Dvoryaninov, G. S. and Prusov, A. V. Estimation of wave parameters in the ITCZ area. *Meteorol. Gidrol.* (1985) (10), 5–13.

19. Citéan, J., Gammas, J. P. and Gouripy, Y. Position de la zone intertropicale de convergence á 28° d'ouest et température de space dans le Golf de Guinée. *Veiller Climatique Satellitaire*. Septembre (1984), Bulletin (3), 2–7.

20. Dvoryaninov, G. S., Prusov, A. V. and Shokurov, M. V. Relationship between the temporal variability of ocean surface temperature and the meridional migrations of the Intratropical Convergence Zone. *Issled. Zemli Kosmosa* (1987) (1), 14–19.

21. Gledzer, E. B. and Obukhov, A. M. Quasibiannual variance as a parametric phenomenon in the tropical system. *Fizika Atmos. Okeana* (1982) **18** (11), 1154–1158.

22. Bubnov, V. A. Seawater temperature variability in the equatorial Atlantic Ocean. *TROPEX-74* (1976) **2**, 62–70.

23. Davydova, E. G., Nedelka, M. I. and Slaba, S. V. Evolution of the Intratropical Convergence Zone as related to the microscopic processes. *TROPEX-74* (1976) **1**, 155–176.

24. FOCAL/SEQUAL Report. *J. Geophys. Res* (1983) **91**, 92.

25. Further Progress in Equatorial Oceanography. A report of the US TOGA Workshop on the Dynamics of the Equatorial Oceans, Honolulu HI, August 11–15, 1986, Eli J. Katz and Janet M. Witte (Eds).

26. Legeckis, R. Long waves in the equatorial Pacific and Atlantic Ocean during 1983. *Ocean-Air Interactions* **1** (1), 1–10.

27. Houghton, R. W. and Colin, C. Thermal structure along 4°W in the Gulf of Guinea during 1983–1984. *J. Geophys. Res.* (1986) **91** (C10), 11, 727–11, 739.

28. Weisberg, R. N. Hickman, J. H., Tang, T. Y. and Weingartner, T. J. Velocity and temperature observations during the seasonal response of the Equatorial Atlantic experiment at 0°, 28°W. *J. Geoph. Res.* (1987) **92** (C5), 5061–5075.

29. Legeckis, R. and Reverdin, G. Long waves in the Equatorial Atlantic Ocean during 1983. *J. Geoph. Res.* (1987) **92** (C3), 2835–2842.

30. Richardson, R. L. and Reverdin, G. Seasonal cycle of velocity in the Atlantic North Equatorial Countercurrent as measured by surface drifters, current meters and ship drifts. *J. Geophys. Res.* (1987) **92** (C4), 3691–3708.

31. Philander, S. G., Hurlin, W. S. and Pacanowski, R. C. Properties of long equatorial waves in models of the seasonal cycle in the Tropical Atlantic and Pacific Oceans. *J. Geophys. Res.* (1986) **91** (C12), 14, 207–14, 211.

Investigations of the Tropical
Atlantic Ocean, pp. 73 – 90
© VSP 1992.

The problem of variability and interrelation of atmospheric and oceanic phenomena in the Tropical Atlantic Ocean: Part II

G. S. DVORYANINOV, V. N. EREMEEV and A. V. PRUSOV

INTRODUCTION

A number of ad hoc programs have been implemented in the Marine Hydrophysical Institute with the purpose of validating the ideas discussed in Part I, and evaluating the disturbances in the Tropical Atlantic Ocean as a response to atmospheric instability (in particular, the ITCZ's kinematics). This part of the study reports on the data derived from analysis of in situ observations conducted in the course of multiple surveys in the Tropical Atlantic Ocean and provided by a series of long-term buoy stations. These investigations have indicated that a sharp temperature front evolving at the time of disturbance development was such that temperature variation in the meridional direction attained 6 to 7 °C over a range of several dozen miles. The marked variability of currents, particularly of the meridional transport, conditioned by instability waves was observed. A cycle of the most energetic waves equalled 29 days, the wavelength $\lambda = 1000$ km, and the phase velocity was west-oriented.

Computations have shown that waves are responsible for large meridional transports of the zonal momentum ($Y = 0.12$ dyn cm^{-2}) and heat ($Q = 80$ W m^{-2}) in the upper 200 m layer. These quantities are comparable with the magnitudes of wind forcing and solar heat fluxes in the region. The flow divergence is evaluated and the exchange coefficients are determined. On average, these are of the order of $\sim 10^{\circ}$ (cm^2 s^{-1}, C s^{-1}) but are essentially dependent on depth and even are likely to change their sign.

The most important problems in the study of large-scale disturbances in the Tropical Atlantic Ocean are outlined in the second section of Part II; the program of further research and appropriate techniques and instrumentation are discussed.

The present paper reports on the results of the analysis of observations carried out in the equatorial Tropical Atlantic Ocean during the period when the ITCZ occupied its northernmost position. The data were collected during Cruise 50 of R/V Mikhail Lomonosov (1988). A number of meridional sections were made in the course of the cruise at the hydrographic test area and four

UDK 551.460:629.78

self-contained buoy stations were occupied. The distance between the sections was 180 miles, and stations were occupied every 30 miles. The southernmost stations were occupied at 2°S, and the northernmost ones at 12°N. Buoys were deployed in the vicinity of 20°30′W. The northernmost buoy was located at 2°N, and the two southernmost ones at 1°N. The distance between buoy 2 and 4 was 160 miles. The pattern of the location of sections and buoys is shown in Fig. 1a.

In fact, the compiled data indicate that a destabilization occurs in the equatorial area of the ocean at this time of year along with large-scale upwellings, unstable waves and sharp temperature fronts. Our objective was to explore these phenomena in the net form; therefore, only those observations were subjected to analysis which pertained to the cases when the phenomena indicated above were stable and well pronounced. However, as the largest period of unstable waves is equal to 30 days, and in the Atlantic Ocean the latter may exist for two or three periods [1, 2], the interval in the data analysis proves equal to one or two periods of these waves as segments encompassing transitional periods associated with the genesis and degradation of the phenomenon must be eliminated. The 30-day long observations made possible the analysis of unstable waves proper and their related effects.

The length of unstable waves is about 1000 km [1, 3]. In order to evaluate simultaneously the wavelengths and the related divergence of heat and momentum fluxes, the buoys were deployed at such distances which made possible solution of both problems. To this end, a series of test calculations were preliminarily performed. The obtained results permitted us to choose the pattern for buoys arrangement. Besides, the area of instability, upwelling, and waves propagate in a westerly direction approximately up to 25–27°W. Therefore, the buoys were deployed near 20°30′W.

Thus, using relatively short records provided by the buoy antenna array, we had, in particular, to evaluate the instability of wave parameters. This task can not be implemented through the use of the traditional techniques of the Blackman–Tjuky type. The most efficient here seem to be the adaptation methods. We applied our own multidimensional variant of the maximum entropy technique [4, 5]. It was shown to have a higher resolution and to yield adequate estimates of the wave parameters from short records (shorter than the considered wave's period) containing much noise and derived from the antenna arrays with a small aperture. Thus, in the considered case, the period at issue equals 30 days and the chosen aperture of the antenna array were large enough to correctly estimate even the maximum period and length of unstable waves. However, for better reliability of the obtained data a special test was additionally carried out.

Buoys taking samples each 10 min at standard depths observed the easterly (u) and northerly (v) velocity components and temperature T. A survey of the test area and buoy-provided data have shown that an upwelling was evolving and equatorially unstable waves were being generated over the period of observations. The best pronounced zonal temperature front was located in the vicinity of 2°N (see Fig. 2), where buoy 1 (B1) was deployed.

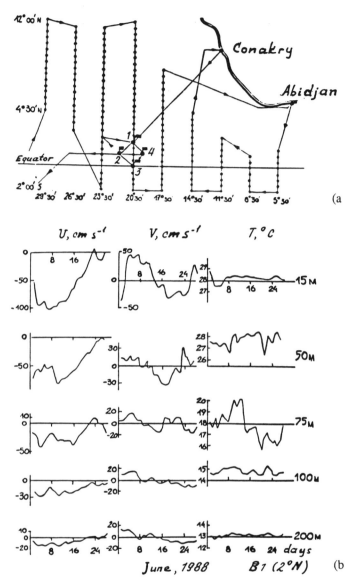

Figure 1. Diagram of the hydrophysical survey and the distribution of buoy stations (a), records of the velocity (u, v) and temperature variations in the upper ocean layer as observed by buoy 1 (b).

Figure 1b shows instances of the variability of the daily means of u, v, and T over the whole registration period (from June 1 to June 30). Even these original data, not subjected to any kind of processing, display a marked variability of the three parameters, the cycle being of the order of a month which becomes

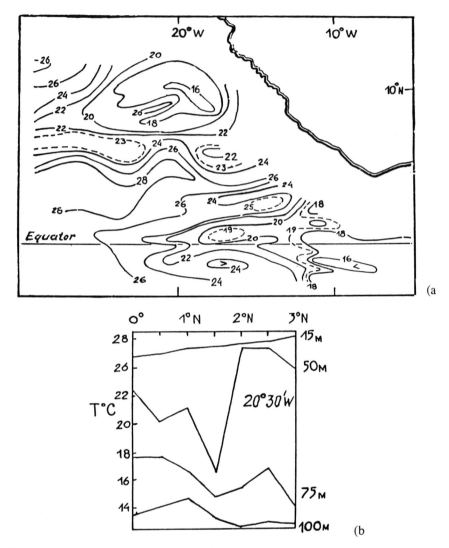

Figure 2. Temperature distribution at a depth of 50 m (a) and meridional variability at $\lambda = 20°30'W$ (b).

particularly evident in the behaviour of the meridional velocity. Aside from this, zonal velocity records imply that the powerful west-oriented flow observed at the initial period of measurement at all depths of the upper 200 m layer then gradually degradated and vanished. It is during this time that the meridional velocity component alternated its sign (it will be appropriately reminded that a similar situation (see Part I) evolving at the time of instability generation is described in ref. [2]). It is also seen that the largest velocities and their variance occurred in the upper ocean layer.

Note that although all these peculiarities of the variability were registered by all buoys, its intensity tends to increase away from the equator which is characteristic of equatorially unstable waves.

It was mentioned above that in the course of the hydrographic survey and acquisition of buoy data, a tongue-like upwelling invaded the equatorial area from the east and reached 25–27°W; accordingly, thermal fronts evolved in the near-equatorial region. This event is demonstrated in Fig. 2. Figure 2a shows the field of temperature isolines at $Z = 50$ m based on the hydrologic survey data. Temperature distribution at the meridional section, $\lambda = 20°30'$W, in the vicinity of the equator where buoys 1 and 3 were deployed, is portrayed in Fig. 2b. The whole section was performed on 7 June, 1988, i.e. the presented data were obtained virtually synchronously.

Wave fluctuations over the observation period are readily visualized in the vector presentation of the horizontal velocity field (Fig. 3a). These data were provided by buoy 1. They demonstrate an existence of wave modulation with a synoptic cycle of the currents' field down to the depth of 500 m. The currents' cyclicity is particularly pronounced within the depth interval of the Equatorial Undercurrent (50–100 m).

To evaluate the energetic wave parameters, buoy data were spectrally analysed using the variant of the multidimensional method of maximum entropy mentioned above. Figure 3b depicts some results recovered from the spectral estimation. The curves in Fig. 3b represent autospectra of the meridional velocity, S_v, for three buoys (S_v for B2 and B4 are identical) at $Z = 15$ m. The diagrams permit singling out of the most energetic peak in a 29-day long period. All spectra also have peaks at the period of ~ 5 days.

To evaluate the magnitude and direction of energetic waves, space–time spectra were computed from the data provided by all buoys. A section of the space–time spectrum (see Fig. 3c) for the most energetic period ($T = 29$ days) indicates that the length of this wave is about 1000 km and the direction is westward.

Coherence spectra, $\varkappa_{uT}^2$, and phase shifts, φ_{uT}, for various depths are presented in Fig. 4. One of the principal deductions ensuing from the distribution of $\varkappa_{uT}^2$ and φ_{uT}^2 is that large values of $\varkappa_{uT}^2$ virtually at all depths occur merely in the proximity of some periods; moreover, in these cases, the values of φ_{uT} roughly equal 0 and π. Hence it follows that basically it is the waves that transport the heat.

Variability of the horizontal velocity components' vertical distribution, depending on the phase of unstable waves, is shown in Fig. 5a (compare with Fig. 1b). The diagrams display a strong variance of zonal and meridional integral flows in the upper 100 m layer. Hence, unstable waves contribute mainly to the dynamical variability of the upper ocean layer. In addition, away from the equator, the variability of flows is more pronounced than in the equatorial area. A meridional integral flow observed by the northern most buoy changed its sign with time, whereas when crossing the equator, it was always directed

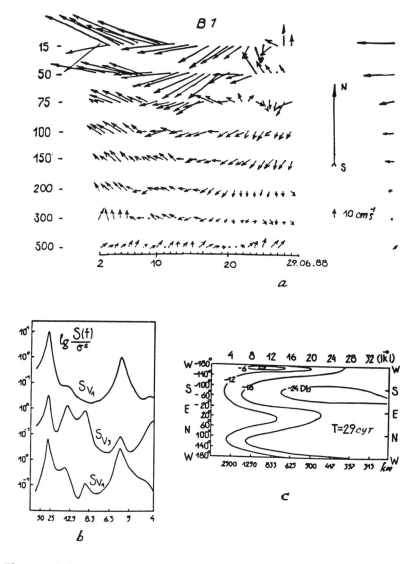

Figure 3. Daily averages of the vector current velocity observed by buoy 1 (a); autospectra
of the meridional velocity for three buoys (b); and a section of the meridional velocity for
$T = 29$ days and $Z = 15$ m (c).

northward. Epures of the time-averaged velocity observations for all the buoys
are shown in the first two diagrams in Fig. 5b (as the upper current recorder
mounted on buoy 2 was not functioning all of the time, velocity epures for
this buoy are not constructed for all depths). The time-averaged meridional
velocity component observed by the equatorial buoy in the entire upper ocean
layer is oriented northward. This means that during the time of existence of

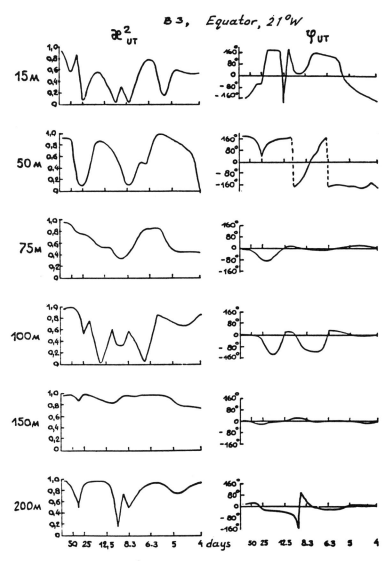

Figure 4. Coherence spectra $\varkappa^2_{uT}$ and phase shifts φ_{uT} between zonal velocity and temperature.

the equatorial upwelling and unstable waves, the integral meridional flow was directed northward and, accordingly, the advective heat transport was southward, as the meridional temperature gradient is positive.

It is seen from the epure of the time-averaged zonal velocity in the equatorial area that the main stream of the Equatorial Undercurrent was at a depth of 75 m and its velocity was 85 cm s^{-1}.

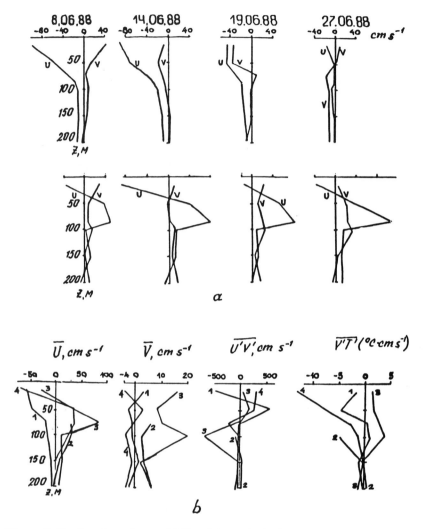

Figure 5. The time-dependent variability of the vertical velocity component distribution for buoys 1 and 2; epures of the mean velocities (u and v) (a) and vertical meridional momentum ($\overline{u'v'}$) and heat ($\overline{v'T'}$) fluxes (b).

The averaged integral meridional flows observed by buoys 1 and 3 thoroughly differ from one another. This implies that vorticity of the averaged integral circulation, i.e. a vertical motion, occurred there.

One of the most important problems relating to the dynamics of the equatorial ocean is the study of the physical mechanism responsible for momentum and heat transport by eddies. The following issues are considered to be of relevance here:

(1) What mechanisms are behind the mixing processes in the equatorial ocean, i.e. what is the contribution from the momentum and heat vortical transport?

(2) Do the vortical transports of heat and momentum essentially contribute to their general budget in the Tropical Atlantic Ocean?

(3) How can the effects of vortical fluxes be parametrized in models of equatorial circulation? Is, in particular, the parametrization involving constant viscosity and thermal conductivity coefficients adequate?

Below, we will make an attempt to assess the contribution from these transports, the appropriate coefficient values and the distribution of these fluxes over depth relying on the *in situ* data.

The last two diagrams in Fig. 5b show distributions of the meridional transport of zonal heat and momentum fluxes over depth for all buoys. The estimates were derived from the original data as averages of $u'v'$ and $v'T'$ over the whole observational period. The vortical meridional fluxes of the momentum prove to be large. In all cases, they are rapidly degradating below the countercurrent's depth, though their distribution in the upper layer varies. Momentum transports across the equator above and below the countercurrent's main stream are negatively-and positively-signed, respectively. Below the current, the transport reverses its sign again. This change in the direction of meridionally-oriented momentum fluxes with $Z = 75$ m is apparently associated with the sign reversal of the derivative over depth from the countercurrent's zonal velocity component $\partial \bar{u} / \partial Z$. In addition, the meridional heat flux across the equator down to 150 m is positive, whilst at the other buoys it is negative. The opposite direction of fluxes are accounted for by the upwelling axis (the area accommodating the coldest waters) being displaced to the north with respect to the equator approximately by 1°, which is readily visualized in Fig. 2a. The values of meridionally-transported thermal fluxes are also large.

The issue is to what extent this momentum and heat exchange depends, on the one hand, upon unstable waves, and on the other hand, upon small (turbulent) mechanisms is important. To elucidate this point, the fluxes were also computed from the estimates of the waves' amplitudes and phases corresponding to the energetic spectral peaks. Their values and distributions over depth for all extra-equatorial buoys proved similar to the respective values of fluxes calculated using the first technique. A discrepancy was an order smaller than the magnitude of the fluxes proper. This allows an inference: transport off the equator was conditioned mainly by equatorially unstable waves. In the equatorial area, an important role belongs likewise to the high-frequency (turbulent) processes.

Besides, the fluxes were computed from the daily averages of the velocity components and temperature. In the non-equatorial area, these proved again close to the values derived through the use of the initial two techniques. This corroborates the predominant role of unstable waves in the meridional mixing process.

Table 1.

z_1 m	$\partial^2 \bar{u}/\partial y^2$ (×10^{-13}cm^{-1}s^{-1})	$\partial(\overline{u'v'})/\partial y$ (×10^{-5}cm s^{-1})	$-\frac{\partial}{\partial y}\overline{u'v'}/\frac{\partial^2 \bar{u}}{\partial y^2}$ (×10^{7}cm^{2}s^{-1})	$\partial^2 \bar{u}/\partial x^2$ (×10^{-13}cm^{-1}s^{-1})	$\partial \bar{u'}^2/\partial x$ (×10^{-5}cm s^{-1})
15	-4.36	-3.60	-8.3	-3.38	3.92
50	-8.21	3.87	4.7	5.22	0.34
75	-7.42	0.24	0.3	0.97	1.58
100	-2.95	2.72	9.2	4.76	-0.75
150	-2.36	-0.04	-0.2	-0.63	-0.17
200	-0.12	-0.09	7.5	0.34	0.03

z_1 m	$-\frac{\partial}{\partial x}\bar{u'}^2/\frac{\partial^2 \bar{u}}{\partial x^2}$ (×10^{7}cm s^{-1})	$\partial^2 \bar{T}/\partial y^2$ (×10^{-14} °C cm^{-2})	$\partial \overline{v'T'}/\partial y$ (×10^{-7} °C s^{-1})	$-\frac{\partial}{\partial x}\overline{v'T'}/\frac{\partial^2 \bar{T}}{\partial y^2}$ (×10^{7}cm^2 s^{-1})
15	11.6	2.02	-0.82	0.4
50	-0.7	—	-2.86	—
75	-16.3	0.13	-1.55	11.9
100	1.6	-0.83	-1.55	-0.8
150	-2.7	-0.72	-0.64	0.1
200	-0.9	0.09	0.36	-4.0

Note that in ref. [6] the appropriate fluxes were evaluated for the Pacific Ocean. Their order of magnitude and peculiarities of distribution over depth are similar to our results. This implies a similarity of mechanisms responsible for vortical meridional transport during the period of equatorial dynamics destabilization.

To estimate the integral effect, integrals for the divergence of vortical momentum fluxes (I) and heat fluxes (Q) in the top 200 m layer were computed. Outside the equatorial area, the value of I proved to be equivalent to the westerly wind force equal to approximately 0.12 dyn sm^{-2}. Thus, by transporting this momentum, the waves considerably affect velocities of the east-oriented currents. As was indicated above (Fig. 5), vortical heat fluxes outside the equatorial area are negative throughout the upper layer, i.e. the heat is transferred by the wave vorticity toward the equator. The vertical integral for these fluxes proves equivalent to the heating of this area by the thermal flux equal to 81 W m^{-2}. Thus, the contribution of waves to the meridional heat and momentum transport is considered large. It is comparable to the impact of the characteristic wind stress and is virtually equivalent to heat fluxes across the ocean surface in this region. This means that instability waves are a central element in the equatorial ocean thermohydrodynamics and undoubtedly must be taken into account in modelling operations.

The vortical horizontal transport in models of the equatorial ocean dynamics is normally parametrized via an introduction of constant viscosity and diffusion coefficients. The values of the order of 10^6–10^7 cm^2 s^{-1} are considered then to be the most appropriate. We have had an opportunity to evaluate horizontal coefficients for heat and momentum exchange from the buoy-provided data. The derived results are listed in Table 1.

It was concluded from the analysis of meridionally-transported momentum fluxes that the integral transport is directed away from the equator. It is seen in the column 3 of Table 1 that, in fact, the total (integral) divergence of the momentum flux over depth is positive, i.e. it is basically north-oriented. At the same time, the integral divergence of the heat flux (column 9 in Table 1) is negative; moreover, it is negative at all levels. Hence, the thermal flux is directed toward the equator. As for the meridional curvature of the mean temperature, $\overline{T}_{yy}$, it modifies from being positive over the countercurrent's main stream to negative beneath it. This is consistent with the fact that the isotherms are convex over the countercurrent's core and are concave beneath it.

The computed viscosity coefficients (A_x, A_y) vary within the range of (0.2–16)$\times 10^7$ cm^2 s^{-1} with the diffusion coefficients $\varkappa_y$ changing within the range of (0.1–11)$\times 10^7$ cm^2 s^{-1}. Moreover, if the coefficient is minimum at the depth of the Equatorial Undercurrent's main stream, $\varkappa_y$, conversely, is maximum here.

Thus, the order of coefficients is, in general, similar to the one used in equatorial ocean dynamics modelling. However, the coefficients are not only likely to change by more than an order over depth but, in addition, alternate their sign. The latter point has been already indicated in refs. [6, 7] for the equatorial ocean.

Analysis of the variability of the high-frequency part of spectrum, compared with the unstable waves, permitted revelation of a regular relationship in the equatorial area between the variation of the dispersion of zonal velocity σ_u^2, calculated from consecutive records and the temperature variability (Fig. 6) in the countercurrent area. Above the countercurrent's main stream, where $\partial \overline{u}/\partial Z > 0$, the values of σ_u^2 are large and alter in counterphase with the temperature variations. At the depth of the current's main stream ($\partial \overline{u}/\partial Z = 0$), the values of σ_u^2 prove small, and there is no regular correlation with the temperature evolution. At $Z > 75$ m, where $\partial \overline{u}/\partial Z < 0$, the values of σ_u^2 become large again but vary in phase with temperature variability. This effect commands special interpretation, though clearly there is a regular relationship between energy variations of the high-frequency variability over depth and the vertical gradient (profile) of the Equatorial Undercurrent zonal velocity component.

In conclusion, it will be noted that tropically-unstable waves were theoretically derived and analysed in refs. [8, 9–12]. Although the models were constructed using different mechanisms for the development of unstable waves and conversion of the of mean motions energy into the wave energy, and there is

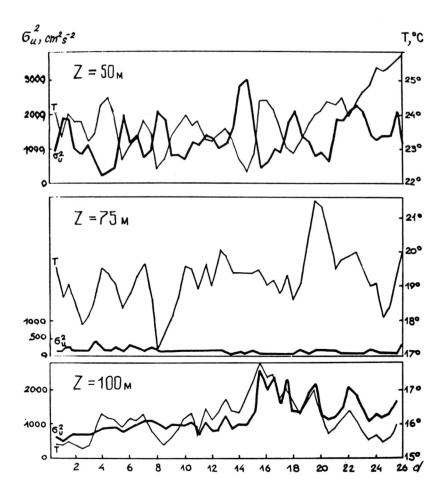

Figure 6. The relationship between dispersion of the high-frequency part of the zonal velocity and temperature fluctuations (at the equator).

as yet no adequate explanation as to how the equatorial ocean dynamics is being destabilized, the models allow an intercomparison of model data with the observations, thereby facilitating a deeper insight into the problem. Modelling results agree with observations in the following aspects: (i) the model, similar to the observations, yields waves propagating westward; (ii) the observed energetic waves and the model ones have similar periods and lengths, which in both cases are most intensive in the upper ocean layer (reaching the lower depth of the Equatorial Undercurrent); (iii) consideration of the wind field's seasonal variability leads to the generation of unstable waves and upwelling during the same seasons when these events actually occur.

However, neither the observations, nor modelling has provided an ultimate response to the major question: what is the actual mechanism behind this phenomenon?

CURRENT SCIENTIFIC PROBLEMS

It follows from the above analysis that the kinematics of the ITCZ and the destabilization of the equatorial ocean dynamics are closely interrelated. Dynamical instability causes a crucial restructuring of the currents field, produces an upwelling and fronts in the ocean, and unstable waves affect the processes of meridional momentum and heat transport. However, there is a wealth of important problems which can be resolved only through the implementation of an appropriate research program. A brief exposé of these problems is given below.

Relationship between the destabilization of the equatorial ocean dynamics and the NECC variability

Observational results and theoretical considerations presented above allow the following deduction: concurrent with the development of a large-scale upwelling and unstable waves in the vicinity of the equator, the wave disturbances also occur in the NECC. These disturbances have a predominant period, which is reminiscent of the period of instability waves, and as the satellite imagery and numerical calculations indicate, have the same wavelengths. However, these fact must be verified against the *in situ* observations using an array of long-term buoys.

It also remains unclear whether unstable waves are localized in the vicinity of the equator and in the NECC area, or the wave disturbances occur over the whole region located between the equator and the NECC northern boundary. This problem is of vital importance as its solution may fill the gap in our knowledge of the mechanism governing Tropical Ocean dynamics, its seasonal perturbations, annual cycle and energetics.

We must also find out whether the NECC wave disturbances represent a response to the generation of equatorially-unstable waves or whether they evolve simultaneously with the latter. Another question: is there a 'local' area where disturbances occur? If the answer is positive, then how large is the temporal delay for the disturbances' reaching the other dynamically significant areas. For example, is the Gulf of Guinea the area where the disturbances are generated with the western boundary layer being the response area, or vice versa?

We have demonstrated in the first section how essential is the role of equatorially-unstable waves in the meridional exchange of zonal momentum and heat. A study of such effects is also necessary in relation to the NECC wave disturbances. This will help us, in particular, to determine whether intensive fluxes of momentum and heat occur only in the sub-equatorial area, or whether they play a central role in the exchange between the sub-equatorial zone and the NECC region.

Finally, the problem of trapping and generation of waves by the North Equatorial Countercurrent is important. It is shown in the second section of Part I that the ITCZ produces synoptic-scale waves which have predominant meridional phase velocity components. The possibility of such a phenomenon has been theoretically proved in refs. [13, 14]. These waves play a major role in the dynamics of the tropical atmosphere as they are highly energetic. Therefore, it would be appropriate to study such effects in the ocean in the context of the NECC's existence and variability. To this end, we should simultaneously investigate the space–time spectrum of wave oscillations inside the NECC and the space–time spectrum of meridional waves outside it but not far from the NECC boundaries. This will permit study of the possible existence of waves of the anti-Kelvin type in the NECC vicinity.

To resolve all these problems, we need to deploy a system of long-term buoys with the antenna array in the sub-equatorial area and in the NECC region; concurrently, a complex survey of the hydrophysical test area and satellite observations of the ocean surface state over the whole period of research should be carried out.

The dependence of vortical fluxes on the structure of mean fields

In order to handle this problem, complex investigations should be carried out at the time when a well-marked destabilization occurs (from May to August) or, conversely, when there is stability. Besides, in order to determine whether an instability, though less pronounced, occurs at the time when the ITCZ occupies its southmost position, appropriate measurements at this time are required. The investigations are expected to reveal which space–time scales are qualitatively and quantitatively important for the dynamics and thermodynamics of the seasonal restructuring, i.e. which processes are continuously contributing to the seasonal variability of the ocean thermohydrodynamics and which of these play the destabilizing role. Hence, an estimation of the space–time variability of energy exchange fluxes between wave (vortical) processes and mean fields is needed. We must find out how large is the contribution of local external effects to the general energy and heat budget and what mechanisms play a central role in the adaptation of the dynamics to variable external effects in this region. A solution of this problem will permit us to gain a deeper insight into the adaptation mechanisms used by numerical models for ocean dynamics and to assess the latter's physical validity. It is also vitally important to determine the adequacy of models and what aspects must be considered to provide authentic model simulations of the variability.

In order to assess the major balance, authentic estimates of the energetic waves' vertical velocity component are required. As was pointed out above, horizontal advective and mixing processes are very intensive in this section of the ocean. This implies that computations of the value of the vertical velocity component from the temperature data will be invalid. At the same time, since the vertical velocities are large during the destabilization period, these can be determined by the law of mass conservation provided that the antenna array for buoy deployment have small bases (not larger than 1°). Hence, the compromise between the necessity to handle energetic waves of about 1000 km long, on the one hand, and the requirement to authentically evaluate mass convergence, on the other hand, imposes constraints on the antenna array parameters. The latter can be chosen on the basis of theoretical calculations. For example, the reliable estimates of the vertical velocity variability in the top 100 m ocean layer equalling 0.1 mm s^{-1}, this being not larger than their real values, can not be recovered from buoy measurements with the distance between the buoys larger than 100 km, unless the difference between the coherent terms of the horizontal current velocity components is equal to approximately 10 cm s^{-1}. It was shown in the previous sections that such differences in velocity values are fairly realistic, and the suggested method of maximum entropy permits true estimates of the largest instability waves (1000 km) using such an antenna array. Consequently, an optimal distance between buoys must be of the order of 100 km.

One of the main reasons warranting further research of instability waves is associated with their marked nonlinearity which produces large vortical fluxes of heat and momentum, in other words, an exchange between 'stationary' dynamics and the disturbances. To evaluate this exchange, authentic information about the spatial variability of the derivatives of these quantities is required, which also calls for small spacing between buoys.

In order to determine the vertical variability of vortical fluxes and vertical velocity, as well as the momentum and heat exchange coefficients and their variation with depth, the data with the vertical step being 15 to 20 m at most are necessary. In addition, to evaluate the integral averaged contribution of the high-frequency part of the spectrum (of the turbulent viscosity and diffusion) and to exchange processes and its dependence upon the wave phase, measurements with discreteness of order of at least several minutes must be carried out. Standard devices designed and manufactured at the Marine Hydrophysical Institute meet this requirement. The same information is needed for determining how the level of turbulence may affect the wave parameters and, conversely, how the turbulence intensity depends on the evolution and phase of the instability wave. To perform such analysis, measurements of duration equivalent to several maximum periods of instability waves are required.

Finally, in the context of this problem, it is important to find out how the meridional wave-induced vortical heat flux $\overline{v'T'}$ directed toward the equator influences the circulation in the meridional vertical plane, i.e. the mixing. Thus, measurements by stationary buoy stations, deployed according to the adopted antenna arrays, conducted simultaneously with satellite observations and hydrophysical surveys will permit a response to a number important issues concerning the Tropical Atlantic Ocean dynamics and its space–time variability.

Small-scale processes and mixing

The problem of turbulent and wave mixing in the ocean is considered to be one of the most important in oceanographic science, much more so in application to the equatorial ocean. This is related to the fact that the results of numerical simulation here are strongly dependent on the type of parametrization of exchange effects [8, 11, 15]. Therefore, in order to develop adequate models, one must be familiar with the small-scale phenomena. Exploration of any events in the near-equatorial and tropical ocean is bound to be inefficient and unrealistic unless it involves a simultaneous complex study of the small-scale processes related to these phenomena.

Presently, there are no data on the relationship between the dynamics of small-scale processes in the Equatorial Atlantic Ocean and the large-scale phenomena, in particular, the instability waves. As has been shown in the first section, the vortical momentum and exchange coefficients are large. However, it remains still unclear how the latter depend on the wave phase and how important is the role of vortical exchange processes in the absence of instability waves. On the other hand, the data presented in Fig. 6 imply a dependence on the energy of high-frequency (with respect to instability waves) phenomena in the mean temperature variability during the existence of instability waves.

Thus, we must find a response to the following questions:

(1) In what manner and to what extent do instability waves affect the small-scale phenomena, including turbulent mixing?

(2) How can the contribution by small-scale events to the averaged equatorial ocean dynamics be parametrized for the purpose of numerical simulation of its seasonal variability?

(3) How and to what extent does destabilization affect the heat budget of the tropical ocean active layer?

MODELLING

As was mentioned in the previous sections, knowledge of input model parameters, such as momentum and heat diffusion coefficients is vital for modelling ocean dynamics. The parameters must be particularly realistic when the ocean's seasonal variability is studied, as the spurious values will filter mesoscale variations, or else will artificially enhance a contribution from the high-frequency part of the spectrum, thereby yielding results which are at variance with reality.

Besides, oceanographic modelling has not provided as yet a response to the main issue: what physical mechanism governs the destabilization and instability waves, though it must be admitted that a number of models based on different assumptions succeed in simulating instability waves and their major characteristics.

Another important issue concerns the mechanisms which are involved in the course of adaptation of the equatorial and tropical ocean dynamics to varying external conditions. What balances are responsible for seasonal and intraseasonal variations?

It has been shown through the use of adaptation models in ref. [15] that mutual adjustment of the ocean dynamics characteristics in the equatorial region is taking place more slowly than outside it. This implies that different mechanisms controlling momentum, energy and heat balance are dominant. However, it remains unclear: (i) how true are the derived data; (ii) what real mechanisms are depicted by models and how large are the errors due to the models' imperfectness; and (iii) how the major balances change in different seasons?

CONCLUSION

The problems discussed above can be resolved only through an intercomparison and combined analysis of the data provided by long-term observations, satellites, and modelling.

As was indicated above, an appropriate program has been initiated at the Marine Hydrophysical Institute several years ago. Some new data related to the study of causes and mechanisms governing the equatorial ocean destabilization are presented in this paper. In the second section of Part II we outlined the most important problems, on which our efforts will be primarily concentrated in future and indicated the basic means and methods required for their solution.

In conclusion, it will noted that the program has been compiled and pursued by the authors of the present paper.

REFERENCES

1. Legeckis, R. Long waves in the equatorial Pacific and Atlantic Oceans during 1983. *Ocean-Air Interactions* (1986) 1, 1–10.

2. Weisberg, R. N., Hickman, J. H., Tang, T. Y. and Weingartner, T. J. Velocity and temperature observations during the seasonal response of the equatorial Atlantic experiment at 0°, 28°W. *J. Geophys. Res.* (1987) 92 (C5), 5061–5075.

3. Legeckis, R. and Reverdin, G. Long waves in the equatorial Atlantic Ocean during 1983. *J. Geophys. Res.* 92 (C3), 2835–2842.

4. Dvoryaninov, G. S., Zhuravlev, V. M. and Prusov, A. V. The method of maximum entropy in multidimensional spectral analysis: theory and testing. Sevastopol: MHI (1987), preprint, part I, 43 p.

5. Dvoryaninov, G. S., Zhuravlev, V. M. and Prusov, A. V. The method of maximum entropy in multidimensional analysis of time series. *Morsk. Gidrofiz. Zh.* (1987) (3), 3–17.

6. Bryden, H. L., Brady, E. C. and Halpern, I. Lateral mixing in the Equatorial Pacific ocean. Further progress in equatorial Oceanography. Rep. of the US TOGA Workshop on the dynamics of the equatorial oceans. Honolulu HI, August 11–15, 1986.

7. Hansen, D. V. and Paul, C. A. Genesis and effects of long waves in the equatorial Pacific. *J. Geophys. Res.* (1984) 89, 10, 431–10, 440.

8. Philander. S. G., Hurlin, S. W. and Pacanowski, R. C. Properties of long equatorial waves in models of the seasonal cycle in the Tropical Atlantic and Pacific oceans. *J. Geophys. Res.* (1986) **91** (C12), 14207–14211.

9. Philander, S. G. Instabilities of zonal equatorial currents. *J. Geophys. Res.* (1978) **83**, 36, 79–36, 82.

10. Cox, M. D. Generation and propagation of 30-day waves in numerical models of the Pacific. *J. Phys. Oceanogr.* (1980) **10**, 1168–1186.

11. Semtner, A. J. and Holland, W. R. Numerical simulations of equatorial circulation. Part I: A basic case in turbulent equilibrium. *J. Phys. Oceanogr.* (1980) **10**, 667–693.

12. Schopf, P. S. and Cane, S. A. On equatorial dynamics, mixed layer physics and sea surface temperature. *J. Phys. Oceanogr.* (1983) **13**, 917–935.

13. Dvoryaninov, G. S. Transport and trapping of the energy of nonadiabatic waves. *Morsk. Gidrofiz. Issled.* (1977) (2), 48–60.

14. Dvoryaninov, G. S. *The Effect of Waves in Boundary Layers of the Atmosphere and Ocean.* Kiev: Nauk. dumka (1982), 175 p.

15. Demin, Yu. L. Hydrodynamical diagnosis of water circulation in the world's ocean. Doctoral dissertation. Sevastopol: MHI (1987), 220 p.

*Investigations of the Tropical
Atlantic Ocean*, pp. 91 – 107
© VSP 1992.

Estimation of the fields of absolute current velocities using inverse methods

V. V. EFIMOV and E. N. SYCHEV

Abstract — Absolute velocities of geostrophic currents are determined by Wunsch's method (WM) and by the Bernoulli function inverse method (BM) using data from four large-scale CTD surveys performed in the Tropical Atlantic Ocean during 1986–1987.
Validation of the different patterns of calculation by Wunsch's method have shown that a one-layer model is most appropriate for test areas with fluid contours. As distinct from Wunsch's method, which yields only one absolute velocity component, normal to the borders of closed cells, the Bernoulli inverse method permits recover of the absolute velocity field, but, on the other hand, proves to be more stringent as to the quality of original data. A new variant of the Bernoulli inverse method is proposed in the paper, which allows a reduction of the influence of synoptic noise on the calculated data.
Distributions of the absolute current velocity component derived through the application of the two methods. Data from four surveys are given.

INTRODUCTION

Observations of oceanic thermohaline fields make possible an application of the so-called inverse methods with the purpose of determining the absolute velocities of currents in the Tropical Atlantic Ocean. It is known that the problems of evaluating the absolute current velocity is vitally important and remains as yet unresolved. Sufficiently accurate direct measurements of small velocities in very deep areas are known to be difficult to perform, and, therefore, remain very scarce. It is doubtful that the situation will drastically change in the near future.

Theoretical calculations of the velocity field, involving global diagnostic modelling, continue to lack sufficient accuracy. Prognostic models using historical data or thermohaline field data recovered from measurement with a sequel-up four-dimensional analysis also call for multiple large-scale field measurements, whereas normally the investigator has to handle the data collected from relatively small surveys restricted in time.

UDK 551.465.5(261)

Other techniques of determining the currents' absolute velocity have become extensively used in recent years, namely, the beta-spiral method, Wunsch's inverse method, one-section method, Bernoulli's inverse method, etc.

The purpose of the present paper is to consider the Wunsch method and the Bernoulli inverse method and the experience accumulated in the course of their application to computing Tropical Atlantic Ocean currents. These methods are based on the laws of conservation and on some assumptions. Normally, the main assumption involves isopycnic motion, in which case the diapycnic mixing is neglected. This assumption is supported by data from multiple salinity measurements performed in the test area, and in particular, by the existence of such a characteristic phenomenon as a subsurface salinity maximum (SSM) nearly all over the Tropical Atlantic Ocean.

In the capacity of original data we used information compiled during four hydrographic surveys, conducted in 1986–1987, of a large-scale trans-Atlantic test area, which encompasses the latitudinal section of the Tropical Atlantic Ocean between 1.5°S and 12°N stretching from South America to West Africa up to the economic zones of the coastal states. The first survey was carried out by three research vessels in April 1986. Other surveys, conducted by two vessels, involved longer periods: the second survey was performed during the period from mid-July to the end of October 1986, the third survey conducted from the beginning of February to the end of March, and the forth survey from the beginning of June to the beginning of August 1987. The CTD-soundings to the depth of 1200 m were performed in the course of surveying, following a meridional tack pattern. The spacing between tacks equalled 1.5°, and the sounding step along the tacks was 0.5°.

THE WUNSCH INVERSE METHOD

To perform calculations using this method, four T, S-data-bases were prepared. Preliminarily, the latter were examined, the gaps interpolated and the measurement data smoothed by applying a three-point sliding mean. The data from the data-bases are located at the joints of a $3 \times 2°$ rectangular grid accommodating the region bounded by 3.5 and 11.5°N and 53.5 and 17.5°W (20.5°W in the case of the 4th survey) and cover virtually the whole of the studied region (Fig. 1), with the exception of the near-equatorial strip of water, where the dynamical method of calculating currents proves ineffective.

Vertically, temperature and salinity observations were implemented at 21 standard levels (0, 10, 20, 30, 50, 75, 100, 125, 150, 200, 250, 300, 400, 500, 600, 700, 800, 900, 1000, 1100, and 1200 m). The retrieved data-bases were complemented with climatic data from Princeton University data-base [1] covering 1300, 1400, 1500, 1750, 2000, 2500, and 3000 m depths with a further step being 1000 m.

As is known, the dynamical method allows determination solely of the vertical current velocity shift versus some reference zero velocity surface (zero surface). In order to determine the full velocity, knowledge of the current's

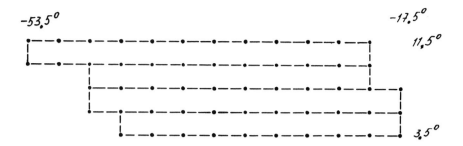

Figure 1. Schematic distribution of the grid joints and the contours of enclosed cells in the test area.

absolute velocity at one of the levels is necessary, e.g. at the zero surface. To determine the current's absolute velocity at the zero surface by the Wunsch inverse method [6], some quantitative characteristics in confined ocean regions, with the sea surface level being constant, are assumed invariable. So, for instance, for any confined oceanic region with the constant sea surface level, a balance between the inflow and outflow of water mass is expected to be maintained, i.e. the next integral over the boundary surface $\mathcal{F}$ must be equal to zero.

$$\iint_{\mathcal{F}} \mathbf{V}\mathbf{n}\,\mathrm{d}\mathcal{F} = 0. \tag{1}$$

Here $\mathbf{V}$ is the vector current velocity at boundary $\mathcal{F}$; $\mathbf{n}$ is the unit vector of the interval normal to $\mathcal{F}$. On the assumption of isopycnic motion of water in the ocean, condition (1) can be written down for one or several layers of water confined between surfaces with constant potential density. In this case, since horizontal and rigid (no-flow) boundaries are not considered, surface $\mathcal{F}$ is constituted by vertical planes of hydrographic sections, i.e. the normal to $\mathcal{F}$ will remain in the horizontal plane, and only horizontal projections of the vector velocity ($\mathbf{V}_\mathrm{H}$) will be used in calculations. In the case of geostrophic motion of water, the horizontal vector velocity can be written down as a sum of two components: the relative vector velocity ($\mathbf{V}_\mathrm{R}$) and the absolute vector velocity ($\mathbf{b}$)

$$\mathbf{V}_\mathrm{H} = \mathbf{V}_\mathrm{R} + \mathbf{b}. \tag{2}$$

The relative vector velocity component normal to the plane of the hydrographic section is determined from the condition of geostrophic-hydrostatic equilibrium of fluid ensuing from the 'thermic wind' equation

$$\mathbf{V}_\mathrm{R}(x, z) = \frac{g}{\rho_0 f} \int_{z_0}^{z} \left(\frac{\partial \rho}{\partial x}\right) \mathrm{d}z, \tag{3}$$

where x is the horizontal coordinate directed along the hydrographic section and the other symbols are conventional.

By unifying equations (1)–(3), we will derive an equation, or a system of equations for several confined ocean regions

$$\left(\iint_{\mathcal{F}} \mathbf{b}\, \mathbf{n}\, d\mathcal{F} = -\iint_{\mathcal{F}} \mathbf{V_R}\, \mathbf{n}\, d\mathcal{F} = -\Gamma \right)_l \qquad l = 1, 2, \ldots, L, \qquad (4)$$

which after having been resolved, will provide the absolute vector velocity. As the current velocity at the hydrographic sections is determined for separate pairs of stations by changing over from the integrals to the sums, we will derive equation (4) in a matrix form

$$A_{L \times N} \cdot B_{N \times 1} = -\Gamma_{L \times 1}. \qquad (5)$$

The system of linear algebraic equation (5) includes a number of unknown quantities (N) equal to the number of nonrecurrent pairs of adjacent hydrographic stations occupied at sections contouring confined ocean regions (cells); and the number of equations (L) is equivalent to the number of confined cells multiplied by the number of layers, for which the conservation equation (1) is written down. In our studies, geometry of the trans-Atlantic experimental region is convenient for obtaining four zonally-aligned cells shown in Fig. 1. Six types of models were examined for these number cells, which differed by the choice of boundary isopycnic surfaces between the layers. With the number of unknown quantities totalling 59 unchanged, the number of equations increased from four to twenty for one-layer and five-layer models, respectively. In all models, the upper and lower horizontal boundaries were represented by 24.80 and 27.87 isopycnic surfaces, which do not intersect with the ocean surface or bottom. The difference between two double-layer models 2a and 2b consists only of the choice of an internal boundary between the layers. Models 3, 4 and 5 were constructed by introducing a respective number of internal boundary surfaces. The list of the boundary isopycnic surfaces and their depths for various models is given in Table 1. The number of unknown quantities is by far superior to the number of equations for any of the six models.

Table 1.
Limits for the data used in models

| | | The boundary isopycnal depth (m) | | | | | |
| | | 19–113 | 50–154 | 150–342 | 275–485 | 857–1005 | 2372–4600 |
Model type	Number of layers	Values of the boundary isopycnals, σ_θ					
1	1	24.8	—	—	—	—	27.87
2a	2	24.8	—	26.8	—	—	27.87
2b	2	24.8	—	—	27.0	—	27.87
3	3	24.8	—	26.8	27.0	—	27.87
4	4	24.8	26.2	26.8	27.0	—	27.87
5	5	24.8	26.2	26.8	27.0	27.4	27.87

This is the main peculiarity and, concurrently, the shortcoming of equation (5) is its underdetermineteness, which leads to its having an infinite number of solutions. In this case, the choice of a proper solution seems impossible unless we introduce some additional solution restrictions. Typically, the so-called normal solution is singled out i.e. the solution which has a minimal modulus, or a minimal norm ($\|\hat{b}\|_2 = \min$). Solution to equation (5), which is minimum by modulus, is unique and can be derived from relation

$$\hat{b}_{N \times 1} = A^+_{N \times L} \cdot (-\Gamma_{L \times 1}), \tag{6}$$

where A^+ is a matrix pseudo-inverse to matrix A. To determine A^+, an expansion by eigenvalues was applied. Table 2 lists examples of the vector Γ components for model 1 determined from four data-bases. Positive vector Γ components denote an inflow of water to a corresponding cell, with negative ones indicating an outflow. It should be noted that virtually in all cases the largest are inflows to the near-equatorial cell 4. To indirectly verify the validity of obtained vectors, Γ, we can use the appropriate values of the compensating mean vertical velocities listed in Table 2. Vertical velocities at the cells' lower boundaries directed upward or downward, correspond to outflows or inflows of water, respectively. Mean vertical current velocities in Table 2 agree with the published data.

Table 2.
Vector Γ components for one-layer models and the mean vertical velocities

Cell	Area	Survey			
l	10^9 m^2	1	2	3	4
		Γ_l, 10^6 m^3 s^{-1}			
1	803	-1.00	0.949	-2.56	-1.95
2	661	-0.611	6.11	-5.10	-2.23
3	738	-3.45	1.83	-12.1	-6.06
4	668	19.6	18.3	8.29	12.9
		$\overline{w}_l$, 10^{-4} cm s^{-1}			
1		1	-1	3	2
2		0.9	-9	8	4
3		5	-2	16	8
4		-29	-27	-12	-19

Figure 2 shows the dependence of the volume transport, modulus and extreme meridional component of the absolute vector current velocity upon the type of model (in other words, upon the number of layers). Hence, it follows that the calculated volume transports across the latitudinal sections are likely to thoroughly transform depending on the number of layers; this is explained by the influence of specific models upon the absolute vector current velocity. The results confirm that both the modulus and the extreme meridional component of the absolute vector current velocity rapidly increase with an increasing number of layers. Besides, solutions for models of one type are notably affected by the choice of the boundaries' position between layers. In Fig. 2 this consideration is exemplified by the results obtained for two-layer models 2a and 2b.

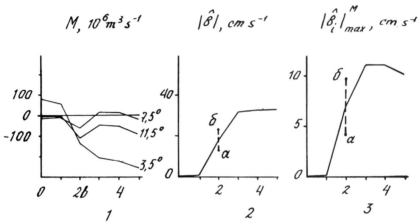

Figure 2. Dependence of the volume transport (1), modulus (2), and extreme meridional component (3) of the absolute current velocity on the type of model for survey 1.

Thus, aside from the uncertainty of solution inherent to the inverse method, because equation (5) is usually underdetermined, we again encounter the uncertainty of solutions, resultant from the choice of isopycnic boundary surfaces and the number of layers. As previously, relying on the principle of minimum modulus solution, we chose a one-layer model, because its solution satisfies conservation conditions of equation (1), and has a minimal modulus, compared with multilayer models.

However, even when a one-layer model is involved, another ambiguity remains associated with the choice of a reference zero surface. Figure 3 exhibits the root-mean-square amplitudes of the absolute vector current velocity and of the relative geostrophic current velocity depending on the position of the zero surface (Z_0) and depth (Z), respectively. It is seen here that the rms amplitude of the absolute current velocity determined by the Wunsch inverse method is relatively small and does not exceed 0.5–0.6 $cm\,s^{-1}$; moreover, these values can be derived if zero surface is near the sea surface or at a depth of about 150 m. It is of interest that the vertical amplitude profile of the absolute current velocity has two minima: a subsurface strongly pronounced minimum at the 75 m level and another smooth minimum at a 800–1200 m depth. The relative geostrophic current velocity amplitude is naturally, zero at a zero level and gradually increases in both directions with distance. Incidentally, in surface ocean layers, where the growth is emphatically large, the relative current

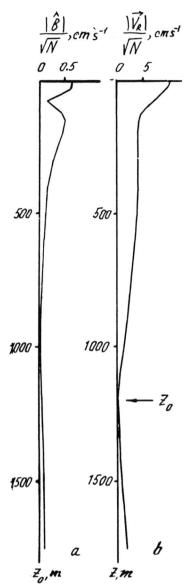

Figure 3. Dependence of the rms amplitudes of the absolute current velocity, determined using a one-layer model (a) and geostrophic velocity (b), on the choice of the zero level and depth for survey 1.

velocity amplitude is an order larger than that of the absolute current velocity. Hence it is seen that in using the Wunsch inverse problem one must be very careful in choosing the reference zero surface, because if randomly chosen, significant fluctuations of the relative current velocity will fail to be compensated by the relatively small absolute current velocity. For example, selecting 0 m as a zero surface will obviously lead to the full current velocity amplitude at the sea surface being considerably smaller than, say, at 1000–1200 m. In various publications, a wide range of depths from 800 to 2000 m and even to 4000 m are applied as a reference surface in the Tropical Atlantic Ocean. However, only a narrow space between 1000 and 1500 m can be considered an optimal choice. The absolute current velocity at these depths is relatively small and does not seriously affect the configuration of the sea surface geostrophic currents. The above arguments support our using the 1200 m level as a zero surface. Computations of the absolute current velocity agree well with the data in ref. [2]. In that paper, the value of the meridional absolute current velocity component, equal to 0.1 cm s^{-1}, was derived at 8°N for the spring season. The corresponding value from our first survey is 0.03–0.04 cm s^{-1}. A discrepancy is accounted for in that a deeper 2000 m level was used in ref. [2] as a reference zero surface, as compared to our case. Figure 3a makes it obvious that even with the 2000 m level involved our model yields a current velocity amplitude of the order of 0.1 cm s^{-1}.

In the framework of the adopted model, the accuracy of calculations mainly depends on the quality of the density field and the errors resultant from the violation of the stationary assumption. To check the latter's validity, we determined the mean quantity ρ_t, which with the 1st and 2nd survey data applied amounted to $\sim 10^{-9}$ kg m^{-3} s^{-1}. Such deviation from the stationary state contributes insignificantly to the total error. The density field quality depends on the errors of measurement, as well as on the natural variability of subgrid scale hydrophysical fields. Errors of temperature, salinity, and pressure measurements by CTD-probes are ± 0.024°C; ± 0.02‰ and 0.5%, respectively. We considered the error of pressure (depth) measurements as a reference error in measurements of temperature and salinity, i.e. $\Delta f = \pm(\partial f/\partial z)\Delta z$, where $\Delta z = z \cdot 0.005$ and $f - T$, or S. The reference error of temperature measurements is nearly zero in the surface layer and decreases from 0.05 to 0.02 °C in the 100–250 m layer and virtually does not change down to 1200 m. The reference error of salinity measurements is also negligeable in the surface layer; below 100 m it changes moderately over depth and remains within 0.001–0.003‰.

The principal cause of the hydrophysical fields variability is associated with a variety of synoptic phenomena, inertial-gravity, tidal, and short-period waves. As is demonstrated in ref. [3], the particularly considerable contribution to the hydrophysical fields' variability is from inertial-gravity waves with periods ranging from one to several days and tidal waves with periods of up to 1 day. Both types of oscillation greatly contribute (up to 15–20 cm s^{-1}) to the observed velocities in the upper ocean layer.

However, due to the peculiarities in their vertical structure (waves with inertial period are those which propagate vertically upward, and tidal wave velocities rapidly decrease in the thermocline below 150–200 m), their contribution to the error of absolute current velocity computations proves to be insignificant. The general level of error was evaluated from the data provided by a multiday station (3.5°N and 46.6°W) occupied during cruise 46 of R/V *Mikhail Lomonosov* in the Tropical Atlantic Ocean. At this station, 54 probings were made hourly from March 20 to 22, 1986. The collected data were averaged and the rms (standard) temperature and salinity deviations over the entire observational period down to the depth of 1200 m were then determined. Temperature deviation in the upper ocean layer is minimal and equals ~ 0.01 °C; in the 100–700 m layer it becomes considerably larger than the measurement error, varying from 0.5 to 0.2 °C; and below 800 m it remains virtually invariable being $\sim 0.07 - 0.08$ °C. Salinity deviation, although to a lesser degree, is also larger than the measurement error. In the upper ocean, it equals $\sim 0.035‰$; in the 100–700 m layer it decreases from 0.06 to 0.02‰; and below 800 m it remains within 0.01–0.02‰. By combining the data from this analysis with the corresponding errors of measurement and reference, we will derive the error of estimation of the field data, depending on depth. In order to determine how large the impact of this error is upon the accuracy of data, a series of calculations involving random noise added to the original and smoothed T, S-data-bases were performed. The Gaussian normally-distributed noise with zero mean and standard deviation equal to the calculated error was then applied. As a result, it turned out that noise corresponding to the level of inaccuracy of our data leads to a rms 30–40% variation of the absolute current velocity. After a preliminary smoothing of the original data, the impact of noise and, accordingly, the magnitude of the total error became twice as small.

Figure 4 shows distributions of the absolute current velocity meridional component at a zero level (1200 m) for four surveys performed in the studied area. Distribution isolines are nearly zonally-oriented here, which seems natural for zonally-oriented cells in the case of the problem being underdetermined. In all cases, the level of the absolute current velocity determined by the Wunsch inverse method exhibits small intensity and rarely attains 0.01 cm s^{-1}. It is seen from the figures that a boundary between the main areas of the current velocity with the north (> 0) and south (< 0) components involved (the 1986 survey data) is located between the 3.5 and 5.5° N sections. During spring of 1987 (the 3rd survey), the areas of differently-oriented current velocities interchanged their positions with the new borderline between them passing along 7.5°N. In summer of 1987 (the 4th survey), this borderline shifted toward 9°N, and another area of currents with south-oriented velocity component emerged below 4.5°N. The intensity of the meridional current velocity fields, as determined from the 1987 survey data, proved slightly lower compared with that of the 1986 fields. Thus, with the cells being zonally-oriented, the Wunsch inverse method permits only zonally-averaged distributions of the meridional current velocity component to be derived. In order to determine the latitudinal current velocity component, the problem must be solved anew, though with different meridionally-oriented configuration cells. This imperfection of the Wunsch inverse method ensues directly from the fact that this method is designed to yield velocity values only at the borders of enclosed cells. Notwithstanding this, it may prove very useful in computations of the averaged meridional or zonal component of the absolute current velocity in areas with fluid contours. One-layer models, therefore, seem to be most convenient.

THE BERNOULLI FUNCTION INVERSE METHOD

In tests, involving the Bernoulli function inverse method, the T, S initial data-bases were arranged following a different pattern. After controlling and interpolating the lacunas, the data were initially smoothed using a three-point sliding mean, first in the vertical, then along the tacks and finally in the latitudinal direction. The studied area and the $3 \times 2°$ rectangular grid were the same. Vertically, the sampling step was equal to 10 m-within 0–1200 m and the climatic data base was not used at all.

It is known that if the current is stationary and the density diffusion is not considered, then each of the three functions is maintained at the streamline [4]: Bernoulli's function in the geostrophic approximation ($B = P + \rho q z$), potential vorticity ($q = f\rho_z/\rho_0$), and potential density (ρ). Hence it follows that with the values of ρ and q fixed, B must be constant and this implies that there exists a single-valued functional relationship $B = B(\rho, q)$ for a stationary flow in the areas with weak mixing. These areas can be also identified through an intercomparison of the q- and S-distributions. In the case of a stationary current and no mixing, these tracer's contours along the isopycnic surfaces will be parallel to one another and to the streamlines. The data recovered from four hydrographic surveys conducted in the Tropical Atlantic Ocean in 1986–1987 were effectively used for computation and construction of the fields of potential vorticity and salinity at various isopycnic surfaces. As a result, it has been stated that distributions of q and S at the isopycnals in the layer $\sigma_\theta = 27 - 27.4$ are quasizonal and may act as indicators of streamlines. This fact may be regarded as one of the premises for application of the Bernoulli inverse method suggested by Killworth for computation of the pressure field and geostrophic velocity. The method is based on the following: the assumption of a single-valued function $B = B(\rho, q)$, the geostrophic and hydrostatic approximations, and the condition of mass and density conservation. Thus we have equations [5]: for geostrophic balance

$$-fv = -P_x/\rho_0, \tag{7}$$
$$fu = -P_y/\rho_0, \tag{8}$$

hydrostatic equilibrium

$$P_z = -q\rho, \tag{9}$$

mass conservation

$$u_x + v_y + w_z = 0, \tag{10}$$

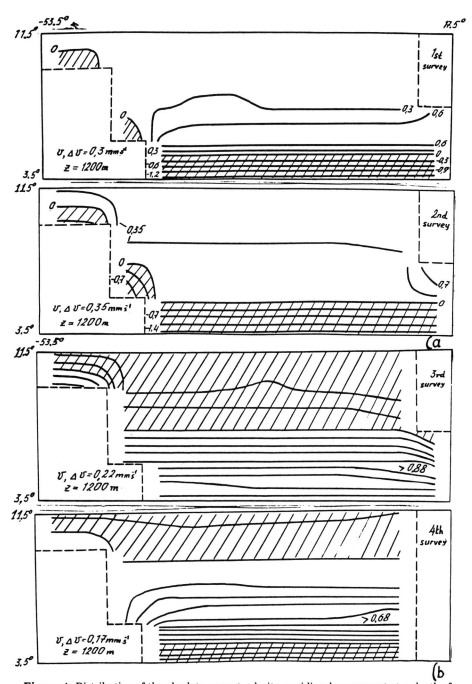

Figure 4. Distribution of the absolute current velocity meridional component at a depth of 1200 m derived using a one-layer model. Hatched sections indicate southerly currents.

and density conservation

$$u\rho_x + v\rho_y + w\rho_z = 0. \tag{11}$$

Here, ρ is the potential density; axes x, y, and z are directed eastward, north-ward, and vertically upward, respectively; the others symbols are conventional.
Through simple transformations we can as well derive

$$uq_x + vq + wq_z = 0, \tag{12}$$
$$uB_x + vB_y + wB_z = 0. \tag{13}$$

Bernoulli's function for the i-th sounding can be determined to the accuracy of some unknown additive constant $\overline{B}_i$:

$$B_i = \overline{B}_i + \int_D^z B_{z_i}\,\mathrm{d}z, \tag{14}$$

where

$$B_{z_i} = zq\rho_{z_i}, \tag{15}$$

and D is some depth, below which the adopted assumptions remain valid.
If, for a pair of soundings i and j, there are depths h_i and h_j, for which

$$\rho_i(h_i) = \rho_j(h_j) \quad \text{and} \quad q_i(h_i) = q_j(h_j), \tag{16}$$

then it follows from $B = B(\rho, q)$

$$B_i(h_i) = B_j(h_j), \tag{17}$$

i.e.

$$\left[\overline{B}_i + \int_D^{h_i} B_{z_i}\,\mathrm{d}z\right] - \left[\overline{B}_j + \int_D^{h_j} B_{z_j}\,\mathrm{d}z\right] = 0, \tag{18}$$

or

$$\overline{B}_i - \overline{B}_j = C_{ij}. \tag{19}$$

Relation (19) yields a linear relationship between unknown quantities $\overline{B}_i$ and $\overline{B}_j$. For a population of pairs of soundings we can derive a population of such linear relationships, which will eventually constitute a system of linear algebraic equations

$$A_{M \times N} \cdot B_{N \times 1} = C_{M \times 1}, \tag{20}$$

which will allow determination of the values of the set of additive constants $\{\overline{B}_i\}$ for the soundings encompassed by equation (20). Formally, system (20) is 'overdetermined', i.e. the number of equations (M) constituting it exceeds the number of unknown quantities (N). However, matrix A of system (20) has an incomplete rank ($r \leqslant N - 1$, $N < M$), as the value of one of the constants $\overline{B}_i$ can be derived through the linear combining of the remainder ($N - 1$) constants. To solve system (20) with involvement of the incomplete rank matrix we again used the method of eigenvalue expansion. The knowledge of $\{\overline{B}_i\}$ allows a construction of the field of Bernoulli's inverse function and then that of the pressure field. The latter is used to determine the fields of the geostrophic velocity's meridional and zonal components by relations (7) and (8).

The presented formalism was applied as a base for respective algorithm, which is designed to establish linear relationships (19) between soundings within the preliminarily chosen range of depths. In our case, this range corresponds to layer $\sigma_\theta =27.0$–27.4. In case of a variety of relationships between stations i and j, the arithmetic mean of C_{ij} is calculated for these.

Table 3 lists a series of specifications, which give a general characteristic of the system of equations (20), with involvement of the data from all surveys considered.

Table 3.
The general characteristic of the BM models

Characteristic	Survey			
	1	2	3	4
Number of unknown quantities (N)	56	56	56	53
Number of equations (M)	175	175	170	165
Total number of relationships	3418	3637	1749	2920
Mean depth for all relationships (m)	656	655	641	669

The number of unknown qualities is equivalent to the number of grid joints, where constants $\overline{B}_i$ are calculated. The number of equations is determined by the number of pair combinations from N-soundings, $C_N^2 = N(N-1)/2$, i.e. $C_{56}^2 = 1540$ and $C_{53}^2 = 1378$. Obviously, this requires a strenuous effort even from powerful computers. The conducted studies have revealed that not all pair combinations are meaningful but only the local ones. We will define local pair combinations as only those pairs of soundings which are within the limits of one box of the coordinate grid. The maximum number of local pair combinations with which a sounding performed in one area may be concerned is 8. For soundings performed at the test area's perimeter, the number of such pair combinations will be smaller, i.e. 3 for the apex of an external angle, 5 for the linear boundary, and 7 for the apex of an internal angle. The remaining so-called nonlocal pair combinations will be discarded as unmeaningful.

Thus, with this approach the number of equations may attain at best the number of local pair combinations (each of these being considered only once), but it may also be smaller if there are not any links between the soundings. For example, we have 5 local pairs in the case of the 3rd survey which are not related to one another. As a result, the number of equations is smaller by 5 units then the maximum one. If any sounding proves to have no relation in any local combination, it must not be considered. Table 3 also shows the total number of relationships for all local pairs. In principle, each relationship may be presented as an individual equation, as has been done in ref. [3]. However, this leads to a bulky system of equations.

We have proposed above to perform relationship averaging over each local pair. On the one hand, such averaging allows a reasonable reduction of the number of equations, and on the other hand — which is no less important — reduces the impact of noise on the calculated data. Mean depths for the whole range of relationships (Table 3) show that the chosen range of depths is not violated.

One of the advantages to the Bernoulli inverse method, whose modification is given above, consists in the relative simplicity of its application and in the fact that the possible presence of data from soundings at different depths does not impose any principal constraints on its application. Besides, in contrast to the traditional dynamical method and other similar techniques, there is no need to introduce any randomly chosen zero surface. The Bernoulli inverse method allows determination of the full pressure field in the test area from which it is possible to calculate, in contrast to the Wunsch inverse method, the distributions of both (meridional and zonal) absolute current velocity components with a finer spatial resolution.

Figures 5 and 6 portray distributions of the current's meridional and zonal velocity components for the four trans-Atlantic surveys. As compared with the smoothed distributions in Fig. 4, those presented in Fig. 5 seem to be more realistic.

The boundaries between the areas accommodating the north and south current velocity components are basically meridionally-oriented; the current velocity amplitudes become as large as several centimeters per second and are more consistent with the current notions about the velocity values at depth. Two 1986 surveys of the areas containing the south current velocity component are situated closer to the edge of the area, and the area accommodating the north current velocity component occupies its central part. In February–March of 1987 (survey 3), the areas with the south current velocity component are mainly concentrated in the western half of the test area, whilst during summer of 1987, they shifted to the east. A predominant direction in the distribution of the zonal current velocity component (Fig. 6) during spring of 1986 (survey 1) is westward, this being the case nearly all over the test area. During the summer–autumn period (survey 2) this area shifts to the north, and a fairly wide section with the east current velocity component evolves in the southern part of the test area. The 1987 distributions (survey 3 and 4) are less regular, being reminiscent of patches with meridionally-oriented boundaries between the areas accommodating the currents of different-orientation. As for

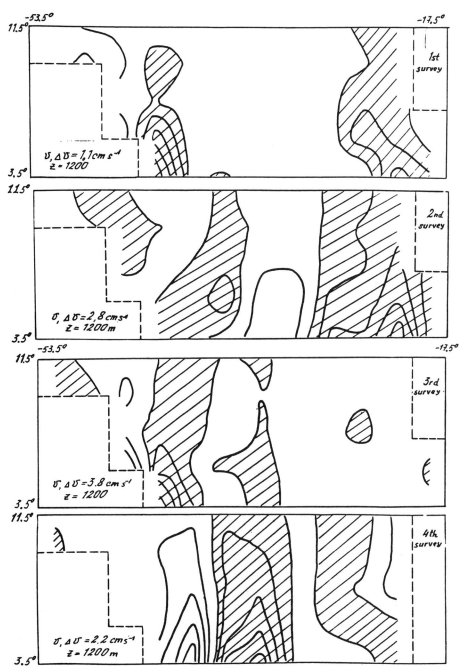

Figure 5. Distribution of the absolute current velocity meridional component at a depth of 1200 m derived by the Bernoulli inverse method. Hatched sections indicate southerly currents.

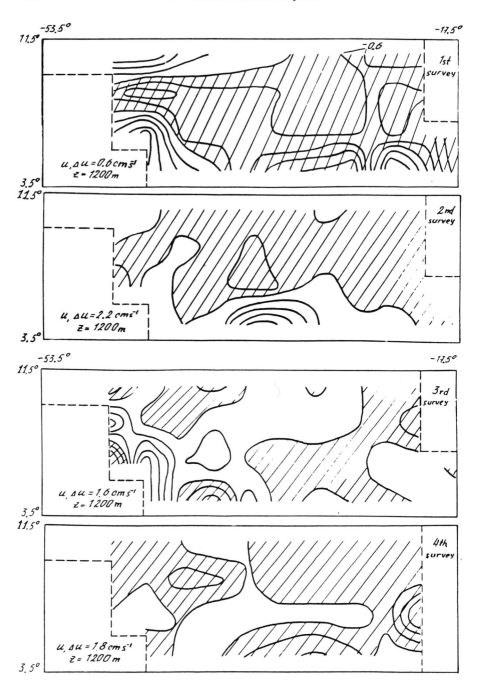

Figure 6. Distribution of the absolute current velocity zonal component at a depth of 1200 m derived by the Bernoulli inverse method. Hatched sections indicate westerly currents.

the accuracy of the absolute current velocity determination using the Bernoulli inverse method it mainly depends, like in the former case, on the quality of the density field and on the errors occurring due to violation of the stationary assumption; in addition, the latter's impact proves better pronounced than in the case of the Wunsch inverse method. This point was taken into account when original *T*, *S*-data-bases, subjected to an additional smoothing in the vertical and horizontal directions, were being prepared. In this case, variations of the absolute current velocity field were reduced by 40–45%, which practically corresponds to the noise level, or even exceeds it by 5–10%.

Intercomparison of the calculations derived through the use of the Bernoulli and Wunsch inverse methods with the zero surface being at the 1200 m depth, has indicated that at upper levels the results are virtually identical. However, essential discrepancies occur at larger depths and the BM field of currents proved to be more intensive.

The available oceanographic publications contain numerous facts pointing out the existence of highly-intensive motion in the interior of the ocean; however, in some cases associated, for instance, with the study of transport at great depths, for verification of the Bernoulli inverse function method it seems appropriate to compare the calculated current velocity with *in situ* observations conducted at large depths, at least at a few points.

In conclusion, it can be noted that owing to the advantages of the Bernoulli inverse function method discussed above, the latter may be successfully applied for calculations of the absolute current velocity at various experimental sites located in tropical and middle-latitude waters. However, a special role is then attributed to the preliminary analysis and preparation of the original thermohaline fields with the purpose of filtering out the nonstationary components.

REFERENCES

1. Levitus, S. and Oort, A. H. Global analysis of oceanographic data. *Bull. Am. Met. Soc.* (1977) **58**, 1270–1284.

2. Roemmich, D. The balance of geostrophic and Ekman transports in the Tropical Atlantic Ocean. *J. Phys. Oceanogr.* (1983) **13**, 1534–1539.

3. Belevich, R. V., Bulgakov, N. P. and Efimov, V. V. Investigations of the tropical energetically-active zone of the Atlantic Ocean. Moscow: Itogi nauki i tekhniki. Atmosphera, okean, kosmos. Programma RAZREZY (1985) **5**, 149–200.

4. Pedlosky, J. *Geophysical Fluid Dynamics.* New York Inc. Springer-Verlag: (1982), 792 p.

5. Killworth, P. D. A Bernoulli inverse method for determining the ocean circulation. *J. Phys. Oceanogr.* (1986) **16**, 2031–2051.

6. Wunsch, C. The North Atlantic general circulation west of 50°W determined by inverse methods. *Rev. Geophys. Space Phys.* (1978) **16**, 583–620.

Investigations of the Tropical Atlantic Ocean, pp. 109 – 122
© VSP 1992.

The mechanisms for seasonal restructuring of long current fields in the North Tropical Atlantic Ocean

G. K. KOROTAEV and G. A. CHEPURIN

Abstract — The mechanisms for seasonal restructuring of the North Equatorial Countercurrent are discussed using theoretical analysis and observational data.

The purpose of the present paper is to describe basic physical mechanisms controlling seasonal variability of the large-scale currents field in the North Tropical Atlantic Ocean in the vicinity of the North Equatorial Countercurrent (NECC). A particular interest in the research in this specific area of the World's ocean is stimulated by the study of heat transport by the ocean from the tropical region toward the high latitudes and of the influence of heat upon short-term climatic fluctuations, in other words, upon the real weather over the European continent.

According to ref. [1], the heat transported by the World's ocean to the north near 20°N accounts for about 70% of the total volume and is equal to $\sim 3.6 \times 10^{15}$ W. Estimates of the heat transport to the north in the Atlantic Ocean range from $\sim 1.55 \times 10^{15}$W across 30°N [2] to $\sim 1.2 \times 10^{15}$W across 25°N [3].

These estimates imply a fairly large contribution from the Atlantic Ocean to the north-oriented thermal transport and stimulate further investigation of the heat redistribution.

A specific feature of the heat transport in the Atlantic Ocean is that it is always oriented northward both in the northern and southern regions [4]. In connection with this, exploration of the North Tropical Atlantic Ocean, where air–sea interaction processes are particularly intensive [5] becomes very important. Oceanic heat fluxes undergo a drastic transformation there, and phenomena occurring in this region effectively contribute to heat transport to the north. We can easily determine the volume exchange between the tropics and subtropics in the North Atlantic Ocean necessary to provide for heat transport, as indicated above. In fact, the north-oriented thermal flux is governed by the relation $Q = c\rho\Delta Tm$, where Q is the heat flux, c is the specific heat capacity, ρ is the density of water, ΔT is the difference in the temperatures of waters transported from the tropical region and the water brought from the high latitude region, and m is the mass of water flowing northward. Substituting the

UDK 551.465.5

above estimates of heat flux and specific heat capacity into this relation, we can readily determine the flow of water mass at issue. If the difference in water temperatures of the outgoing and incoming flows is 5 °C, which is characteristic of horizontal circulation cells, it amounts to about 75 Sv; when the vertical circulation is involved, the difference of temperature being 20–25 °C, the volume equals ~ 10–20 Sv. In the first case, observations of the current are fairly accurate, whereas in the second case, the accuracy reaches the lowest margin.

Investigations conducted in 1982–1984 revealed the presence of strong seasonal variability and regularity in the behaviour of geostrophical current field in that region of the ocean.

Conditionally, two types of circulation may be singled out. The first type of circulation is characteristic of the autumn–winter season, when the NECC evolves in the form of a powerful jet current separating the cyclonic and anticyclonic gyres. Volume transport by the NECC at this time of the year attains 36 Sv and the observed velocities are as large as $2 \, \mathrm{m \, s^{-1}}$. The current is readily observable at the sea surface, as the data on ship and surface float drifts indicate [6]. The NECC jet at this time propagates approximately between 4 and 5°N. Incidentally, it can be appropriately noted that initially it is documented in the southern section of the test area and then gradually shifts to the north. Similar behaviour of the NECC is described in ref. [6], where historical data on surface currents recovered from ship observations are analyzed. The second type of circulation is inherent in the spring–summer period. The NECC is not very intensive at this time of the year and represents a combination of a variety of relatively weak jets. Much noise in the pattern of the large-scale circulation caused is by synoptic–scale phenomena. Analysis of the calculations of the dynamical topography in the north western Tropical Atlantic Ocean indicates that there is no powerful (capable of thermal transport equal to $1.5 \times 10^{15} \mathrm{W}$) geostrophic current all along the South American coast. The existing current readily vizualized from the drifts of ships heading for Brazil, Guiana and Surinam ports represents a purely drift flow.

Let us briefly concentrate on the drift transport, which is always directed northward. Its seasonal variability in transition across the Atlantic Ocean, determined from the climatic wind field data [7] (Fig. 1) range from 28 Sv in winter to ~ 10 Sv in summer. Such transport may close the appropriate vertical circulation cell and provide the necessary export of heat. However, geostrophical transport in the upper 1000 m layer of the tropical ocean is oriented southward and is sufficiently intensive. Analysis of the hydrographic data [8] has shown that seasonal variability in the tropical region is predominant. All this calls for an in-depth study of the physical mechanisms responsible for the seasonal variability of large-scale currents all over the North Tropical Atlantic Ocean.

In order to explore the Seasonal cycle of large-scale circulation in the North Tropical Atlantic Ocean, we have carried out investigations in five trans–Atlantic areas.

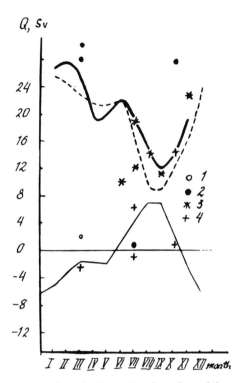

Figure 1. Volume transports through the northern boundary of the trans-Atlantic test area in the Tropical Atlantic Ocean: (1) indicates the geostrophic transport as observed in the test area; (2) is the geostrophic and drift transport in the Tropical Atlantic; (3) depicts the drift transport reconstructed from the charts of wind field; (4) is the regular component of the geostrophic transport. The dashed line indicates the drift transport recovered from the climatic wind field data, the solid line indicates the geostrophic transport based on model simulations with involvement of wind data.

Firstly, it should be emphasized that circulation is essentially zonal at the upper ocean levels and meridional in greater depths. Two specific types of water circulation corresponding to various seasons were observed in the north-western Tropical Atlantic Ocean and are characteristic of the entire circulation in the basin for depths attaining 350 m. Initially, the NECC is well pronounced as a prodigeously meandering stream. In the eastern part of the test area the NECC bifurcates, with one branch travelling to the south and the other one to the north along the African coast. The second branch is not observable as a continuous jet and is only constituted by a number of weak flows with a pre-vailing zonally-oriented transport. Characteristically, seasonal variability also encompasses deeper waters, where it manifests itself in the form of temporally-variable meridional flows, regularly alternating their direction of propagation. To analyze the possible mechanisms for heat transport, geostrophic flows in the upper 800 m layer (zero of the first baroclinic mode in that region is located at a depth of 1000 m) were determined for each test area. The acquired data on volume transport by geostrophic currents across the northern border of the test area are presented in Fig. 1.

As was expected, the geostrophic transport amplitude is comparable with that of the drift transport. The summary transport is directed to the south. A large scatter of data in the summer season, on the one hand, may apparently be accounted for by computational inaccuracies and, on the other hand, by manifestations of the interannual variability. Figure 2 shows zonal distribution of the meridional flow in the upper 800 m layer at the test area's northern boundary for the four surveys. It also demonstrates the transport decomposition into the harmonic and noise components for each survey. It is readily visualized here that the regular current velocity component accounts for 50% of the dispersion of variability in water mass transport along the northern border of the area. However, the regular component's contribution to the total transport is modest (Fig. 1), as the test area accommodates nearly all wavelengths. The wavelength thus determined varies from 2000 to 3500 km. Hence, it may be supposed that the observed geostrophic transports are coupled with

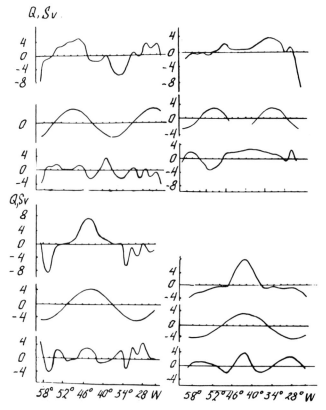

Figure 2. Longitudinal distribution of the longitudinal volume transport through the northern boundary of the test area for four surveys; the regular and noise components are presented in the lower diagrams.

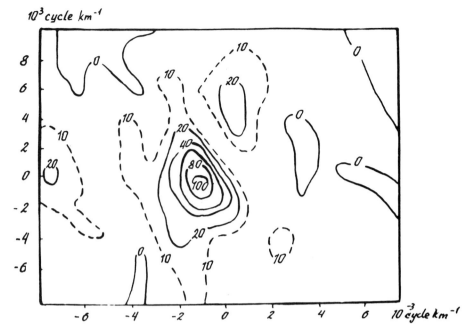

Figure 3. A section of the space–time spectrum of density field fluctuations of annual cyclicity for 100 m depth.

the irregularity of the annual course, as a 3000 km wavelength is close to the Rossby wave of annual cyclicity for this region.

For a better insight into the large-scale wave phenomena in the Tropical Atlantic Ocean, a space–time spectral analysis of the density field in the studied area was performed. Space–time spectra of the density fields fluctuations at various depths and of the related space–time spectra for various pairs of levels were evaluated following the method for field data treatment in ref. [9] and applied for processing POLYMODE data [10].

To correctly interpolate the estimates thus obtained, analysis of the antenna array space–time window formed by the stations occupied in the studied area is of particular significance. Variability, with scales ranging from ~ 400 to ~ 5000 km in the zonal direction and from ~ 120 to ~ 1700 km in the meridional one, can be effectively resolved in the studied area. In fact, the events with a duration varying from several to 18 months were resolved. Estimation was performed for all standard levels confined between 50 and 1200 m. Above 50 m the mixed layer was located which has its own specific dynamics.

Analysis of the sections of space–time spectra at the annual period for various ocean depths allowed us to identify two ocean layers where the spectra are essentially different. Two maximums in the annual cycle are well pronounced in the upper 300 m layer. The space–time section of the spectrum for a 100 m depth corresponding to the pycnocline's mean depth in the test area is shown in Fig. 3. Two maxima are readily visualized here: the first maximum corresponds to a 4600 km long wave propagating nearly zonally in a westerly direction with a phase velocity of $\sim$ 12.5 km d^{-1}. Another maximum corresponds to a 1400 km long wave travelling northward with a speed of 4 km d^{-1}.

Below 300 m, the position and number of spectral maxima are essentially different (Fig. 4). As at the upper levels, the major energetic maximum stands out, which corresponds to a 4600 km wave propagating westward with a phase speed of 12.5 km d^{-1}. However, the second maximum (Fig. 3) below 300 m is not traceable here. Instead, another two maxima emerge (Fig. 4), each attaining about 80% of the magnitude of the first maximum. One of these (Fig. 4) corresponds to a 1800 km long wave propagating in a westerly direction, while the other maximum (Fig. 4) corresponds to a 1700 km long wave travelling eastward. The position of the two maxima with respect to one another, their similar scale and roughly identical intensities, allow us to infer the presence of a standing wave in the investigated area.

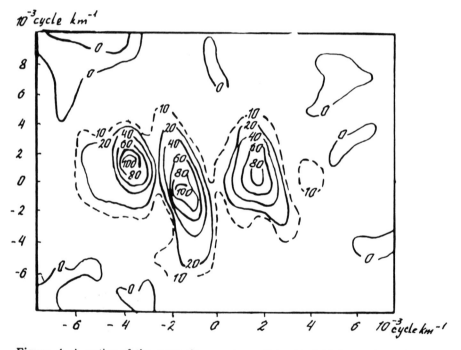

Figure 4. A section of the space–time spectrum of density field fluctuations of annual cyclicity for 400 m depth.

Let us consider in more detail the physical nature of the observed maxima in the space–time spectrum. Firstly, consider the vertical structure of the free wave motion of planetary scale. It is described by the solution to the following eigenvalue problem

$$(\omega - Uk - Vl)\left\{\frac{\partial}{\partial z}\left(\frac{f^2}{N^2}\frac{\partial\varphi}{\partial z}\right) - (k^2 + l^2)\varphi\right\}$$

$$- \left\{k\left[\beta - \frac{\partial}{\partial z}\left(\frac{f^2}{N^2}\frac{\partial U}{\partial z}\right)\right] - l\left[\frac{\partial}{\partial z}\left(\frac{f^2}{N^2}\frac{\partial V}{\partial z}\right)\right]\right\} = 0,$$

(1)

$$z = 0, \qquad \frac{\partial\varphi}{\partial z} = \frac{l\frac{\partial V}{\partial z} - k\frac{\partial U}{\partial z}}{(\omega - Uk - Vl)},$$

(2)

$$z = H\frac{\partial\varphi}{\partial z} = \frac{\left(\frac{\partial V}{\partial z} - \frac{N^2}{f}\frac{\partial H}{\partial x}\right)l - \left(\frac{\partial U}{\partial z} - \frac{N^2}{f}\frac{\partial H}{\partial y}\right)k}{(\omega - Uk - Vl)},$$

(3)

where $\varphi(z)$ is the function describing vertical dependence of the velocity field in a wave derived via the solution of the quasigeostrophic equation for potential eddy conservation using the variables separation method, $U(z)$ and $V(z)$ are the zonal and meridional mean current velocity components, respectively, $N(z)$ is the mean profile of the Brunt–Väisälä frequency, H, $\partial H/\partial x$ and $\partial H/\partial y$ are the mean ocean depth and mean bottom inclination, respectively, assumed constant in conformity with the geometrical optics approximation, f is the Coriolis parameter, β is the latitudinal variability of the Coriolis parameter, k and l are the zonal and meridional wave vector components, respectively, and ω is the wave frequency assumed positive.

It is seen from the profiles of the geostrophic current velocity components [11] that no critical layers evolved in the ocean interiors for the considered space–time maxima. Hence, it follows that the solution to equations (1)–(3) involves a complete discrete set of orthogonal functions $\varphi_n(z)$, which corresponds to various natural mode oscillations. Let us impose the vector components k and l corresponding to a 4600 km long zonal wave propagating westward. We will then find in the course of solving equations (1)–(3) that the first baroclinic mode frequency of the planetary Rossby wave equals 334 days and the phase velocity of its propagation is 13.77 km d^{-1}. The vertical distribution of the velocity and density field fluctuations in such waves are indicated in Fig. 5 by the 1st and 2nd maxima curves, respectively.

It is seen here that density fluctuations remain coherent up to a depth of 1000 m, and that the oscillation maximum is located in the main pycnocline at a depth of 75 m. By comparing the derived solution with the vertical distribution of the amplitude and phase of the first maximum (Fig. 3), as well, its climatic characteristics with the calculated ones, it can be argued that the presence of maximum 1 in the space–time spectral estimates is motivated by the transition of the free Rossby wave of seasonal cyclicity in the studied trans-Atlantic area.

To account for the peculiarities of maximum 2, we will focus on three-dimensional planetary waves travelling in the vertically heterogeneous ocean. As in the previous case, the vertical distribution will be governed by equation (1). In

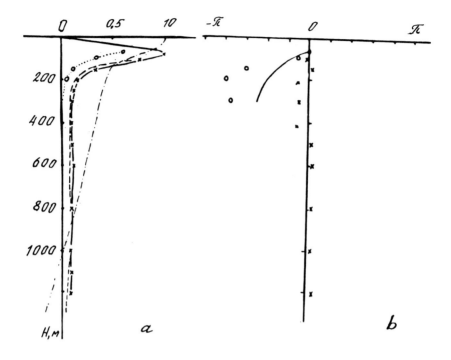

Figure 5. Vertical distribution of the amplitudinal (a) and phase (b) characteristics of spectral maxima. The dash-and-dot line indicates maximum 1; the dotted line indicates maximum 2; and the dashed line indicates the solution to the dynamical problem.

accordance with te WKB-approximation, we will seek a solution to equation (1) in the form

$$\varphi(z) = A(z)c^{i\theta(z)}.$$

Assuming then that the amplitude $A(z)$ of the length of a wave is changing slowly, we will find a dispersive relation for vertically–propagating planetary waves

$$\omega = U(z)k + V(z)l - \frac{\beta k + k\frac{\partial}{\partial z}\left(\frac{f^2}{N^2}\frac{\partial U}{\partial z}\right) - l\frac{\partial}{\partial z}\left(\frac{f^2}{N^2}\frac{\partial V}{\partial z}\right)}{k^2 + l^2 + m^2\frac{f^2}{N^2}}, \qquad (4)$$

where m is the vertical wavenumber.

Energy diffusion in the vertical will take place with the group velocity being

$$c_{gz} = \frac{\partial \omega}{\partial m} = \frac{2f^2 m\left[\left(\beta + \frac{\partial}{\partial z}\left(\frac{f^2}{N^2}\frac{\partial U}{\partial z}\right)k - l\left(\frac{\partial}{\partial z}\left(\frac{f^2}{N^2}\frac{\partial V}{\partial z}\right)\right)\right]}{N^2\left[k^2 + l^2 + m^2\frac{f^2}{N^2}\right]^2}. \qquad (5)$$

The presence of maximum 2 in the space–time spectral estimates may be accounted for by the three-dimensional planetary waves generated on the sea surface by seasonal displacement of the calm area. Qualitative discrepancy between the calculated delay with the space–time estimates (Fig. 5) may be explained by the effect of the WKB-approximation applied for solving equation (1).

The presence of spectral maximums 3 and 4 at great depths may be apparently related to the specific bottom topography in this region of the world's ocean. The characteristic scale of the bottom features, related to the Demerrer elevation in the West Atlantic, the Mid-Atlantic ridge in the central ocean, and the Sierra Leone elevation in the East Atlantic Ocean, exactly corresponds to the spatial scales of maximums 3 and 4. Their manifesting themselves at seasonal periods implies that annual oscillations reach the seabed. We may hypothesize that these maxima occur as a result of interaction between the free Rossby wave of annual cycli city and the bottom features.

Thus, the Tropical Atlantic Ocean response to wind speed seasonal fluctuation may be presented in the form of forced three-dimensional planetary waves generated by the calm area shifts, which transfer perturbations on the sea surface into the interiors of the ocean with an appropriate delay, and as a free oceanic response in the form of a long wave, generated by the eastern shore (as a reaction to variations of the forcing factor) propagating to the west. It is this long wave that provides the seasonal restructuring of the Tropical Atlantic Ocean circulation.

Let us consider a simple model for the currents' response in the tropical section of the North Atlantic Ocean to seasonal variability of the wind field.

It may be supposed that a physical mechanism responsible for restructuring of currents and their adaptation to variable external conditions is associated with the wave processes. Seasonal variations of the tangential wind stress at the sea surface contribute to the generation of waves of annual cyclicity. Generally speaking, several types of waves are expected to be generated in the Tropical Atlantic Ocean, i.e. Rossby waves, Kelvin waves, and the mixed Yanai waves. However, the Yanai and Kelvin waves display a wavelike mode of propagation and are significant merely in the narrow subequatorial area. These waves greatly influence the NECC's seasonal restructuring and must be considered in contructing a closed system of the tropical ocean circulation. In the remaining part of the Tropical Atlantic Ocean the seasonal restructuring of currents is dependent on the propagation of planetary Rossby waves. With the above in mind, and without making any attempt at constructing a complete pattern of the Tropical Atlantic Ocean circulation, we will concentrate only on the study of Rossby waves of annual cyclicity contributing to the seasonal evolution of the NECC. As is seen from the dispersive relation for the latter waves

$$\omega = -\frac{\beta k}{k^2 + l^2 + 1/R_n^2}, \tag{6}$$

where k and l are the zonal and meridional wave vector components, respectively, and R_n is the deformation redius of the nth mode of oscillation; it should be added that two waves of different length have the same frequency. In the case of tropical waves of annual cyclicity, one of these waves is shorter than ~ 5 km, which is far inferior to the characteristic spatial extent of large-scale circulation. Besides, for short waves, the dissipation phenomena are of great importance. In view of this, in further investigations of the NECC's seasonal variability we will take into account only the long Rossby waves and will restrict ourselves to the consideration of the lower modes.

The phase velocity of long planetary Rossby waves is equal to βR_n^2. The deformation radius of the barotropic wave is about 2000 km. Hence, latter crosses the Atlantic Ocean within the space of a few days, and the barotropic response to the wind speed seasonal fluctuations may be considered instantaneous. The velocity of propagation of the Rossby wave's first baroclinic mode is approximately 5 cm s^{-1}, which is consistent with the seasonal variability of the oceanographic characteristics in this region of the Atlantic Ocean.

Thus, the ocean's response to the variable wind force is constituted by the barotropic wave, which instantaneously propagates over the entire basin providing a Sverdrup balance, and by the slow baroclinic waves responsible for seasonal restructuring of the current field with an appropriate temporal delay. As the deformation radii in the Tropical Atlantic Ocean tend to change considerably from 40 km at 20°N to ~ 250 km at the equator, a delay in the oceanic response to the shifting of the calm area and to the change in the intensity of trade winds depends on the latitude. It is obvious that the oceanic response will be different, if the calm area shifts to the south and then re-occupies its northernmost position [11].

The above speculations inevitably reduce the scope of our research to the study of the influence of wave phenomena occurring due to the shifting of the calm tropical area, upon the seasonal evolution of the North Equatorial Countercurrent.

The direct estimation of terms in the potential eddy conservation equation involving the historical Tropical Atlantic Ocean wind and hydrological data performed in ref. [10] has shown that far away from the shore, only the linear equation for potential eddy conservation is valid, which considers the first baroclinic mode of oscillations. Therefore, to describe oceanic motions, we will use the following equation

$$-\frac{1}{R_n^2}\frac{\partial \psi}{\partial t} + r\frac{\partial^2 \psi}{\partial x^2} + \beta\frac{\partial \psi}{\partial x} = -\frac{\partial \tau_x}{\partial y}, \tag{7}$$

which considers only one baroclinic mode.

The first term in the equation describes the temporal vorticity variance conditioned by baroclinic wave propagation. Here, $R_n(y)$ is the nth mode deformation radius and $\psi(x, y, t)$ is the stream function. The second term denotes vorticity dissipation due to bottom friction in the Rayleigh form. The bottom friction should be necessarily taken into account in order to set a boundary condition at the western border of the basin and to describe the western boundary currents. The third term indicates the conventional vorticity evolution associated with the latitudinal variability of the Coriolis parameter. The right-hand part of equation (7) describes vorticity generation by the tangential wind stress.

Equation (7) governs long nondispersive Rossby waves travelling with a speed of βR_n^2. Note that the solution to equation (7) depends parametrically on latitude, as the deformation radius $R_n(y)$ and the tangential wind stress $\tau_x(y)$ depend on y. The boundary conditions for ψ are as follows

$$\psi(0, y, t) = \psi(L, y, t) = 0 \tag{8}$$

The solution to equation (7) with the boundary conditions of (8) can be written down for a arbitrary wind distribution

$$\psi(x, y, t) = R_n^2(y) \int\limits_{\frac{X - L}{\beta R_n^2} + t}^{t} \frac{\partial \tau_x}{\partial y} \, d\xi - R_n^2(y) e^{-\frac{\beta}{r} x} \int\limits_{-\frac{L}{\beta R^2} + t}^{t} \frac{\partial \tau_x}{\partial y} \, dr\xi. \tag{9}$$

Deformation radii of the corresponding modes were determined by solving the eigenvalue problem

$$\frac{\partial}{\partial z} \left(\frac{t^2}{N^2} \frac{\partial \psi}{\partial z} \right) - \frac{1}{R^2} \psi = 0, \tag{10}$$

$$\frac{\partial \psi}{\partial z} = 0, \qquad z = 0, H.$$

The Brunt–Väisälä frequency distribution was defined from the Princeton University climatic data-bases. Computations were performed for the real geometry of the shore.

Firstly, computations were performed for model wind speeds prescribed by the relation

$$\tau_x = -0,6 \left[1 - e^{-(y - 500(0,1x + \sin \omega t))^2 / 250} \right], \tag{11}$$

where y is the distance from the equator given in kilometers and $\omega = 2\pi/T$ is the frequency corresponding to the annual harmonic ($T = 365$ days). Such tangential wind stress distribution corresponds to the width of the calm ocean area equal to about 500 km and to its displacement away from the equator as far as 12°N. The southernmost position of the calm area was documented in April.

The calculated data are portrayed in Fig. 6. Firstly, it should be noted that generation of the wave processes does not lead to the vanishing of the North Equatorial Countercurrent. However, the latter's intensity undergoes appreciable changes depending on the season. The NECC has been observed to be most intensive when it occupies its southernmost position and, conversely,

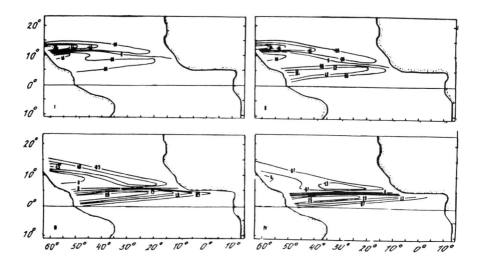

Figure 6. Streamfunction of the solution to problems (8)–(9) for wind [12].

becomes very weak during winter when it shifts northward. Fluctuations of the North Equatorial Countercurrent's intensity may be accounted for by two reasons. Firstly, by the West African coast's configuration (the wind acceleration area in the north section of the basin is 1.5-fold smaller then in the south); and secondly, by the reduction of the deformation radius with latitude, which makes the velocity of propagation and length of waves of annual cyclicity smaller in the northern region. Reduction of the phase velocity of long waves with latitude leads to the widening of the NECC and degradation.

Shifting of the NECC, like the displacement of the calm area, occurs within on ocean region bounded by the equator and 12°N. However, it does not exhibit any fluctuation. The NECC is generated near the equator at the end of spring. At this time of the year, there still exists a degradating branch of the countercurrent in the north near ∼ 12°N. The northern branch then vanishes and the southern one becomes more intensive, attaining its maximum at the end of summer. The NECC then shifts north, where it terminates its existence during the following spring, after which the annual cycle repeats itself. A delay in the countercurrent's shifting near the western shore with respect to the displacement of the calm area ranges from 3 to 6 months and depends on the latitude. It is maximum in the northern area and minimum in the southern one.

Such seasonal evolution of the NECC may be explained by the fluctuation of the phase velocity of baroclinic Rossby waves with latitude. It is maximum near the equator, therefore, perturbations of the current generated near the eastern shore have enough time to rapidly cross the ocean and produce a response to the changing wind force. A delay in shifting of the countercurrent's stream with respect to the calm area is minimum in this region. When the calm zone shifts to the north, the velocity of wave propagation tends to decrease, which results in the lowering of the countercurrent's intensity and in the delay in its shifting with respect to the calm area. When the latter starts shifting southward, the NECC fails to evolve due to the delay in baroclinic response, untill the calm area does not immobilize at its southmost position.

The presented model permits estimation of a contribution of the regular annual wave to water mass transport across the area to the north directly from the known climatic wind field data. The appropriate procedure is as follows. Firstly, the stream function horizontal component is determined using climatic wind field data (11) and the simple model described above. Furthermore, its amplitude is determined from the vertical eigenfunction for the first baroclinic mode and from spectral estimates of the density field fluctuations applying the root-mean-square technique. The density field then formed by the first baroclinic mode is reconstructed and the transport volume is calculated in a conventional manner. (We could neglect spectral estimation and assess the first baroclinic mode amplitude directly from the projection of the wind field onto the first mode. This method is expected, however, to yield results, which are less comparable with direct measurements owing to the relatively large interannual amplitudinal variability of the wind and its relatively small spatial variability).

Transport estimates derived in this fashion are given in Fig. 1. These display a good qualitative agreement with the transport estimates derived through harmonic analysis. Here, we can trace the onset of intensification degradation phases, as well as the reversal of the sign of water mass transport.

Thus the dynamics of the seasonal restructuring of the Tropical Atlantic Ocean currents field appears as follows. The Intratropical Convergence Zone shifting to the north and the intensification of southerly trade winds in summer contribute to the generation of planetary waves. In the southern part of the studied tropical area, rapid wave propagation stimulates the southern anticyclonic gyre's fast adaptation to the modified wind field. During this period the thermocline is observed to deepen in the western part of the anticyclonic gyre with a huge amount of warm water being accumulated in it. Concurrently, the NECC emerges and intensifies. It is traceable as one continuous flow crossing the entire Tropical Atlantic Ocean at depths reaching ~ 350 m. In the eastern part of the studied area, the NECC bifurcates. One branch of it travels to the south around the Sierra Leone elevation, and another branch propagates northward in the vicinity of the African coast. Reduction of the velocity of planetary waves away from the equator is conducive to a large delay in the response of the northern cyclonic gyre, to the changed wind direction, as compared to the southern anticyclonic gyre, and to the different oceanic response to the ITCZ's shifting north or south. In the latter case, owing to the delayed oceanic response, the NECC continues to shift to the north in the western part of the area up to 11–12°N, simultaneously becoming less intensive.

In spring, the NECC's intensity is minimum and the countercurrent ceases to exist as one continuous flow crossing the Atlantic Ocean. During this period, the countercurrent's degradation is obviously spurred by the instability of the newly–formed front.

In the eastern part of the area, a relatively weak countercurrent jet continues flowing equatorward by virtue of the local wind direction.

The NECC degradation in the western part of the investigated area during spring facilitates penetration of warm waters from the southern anticyclonic gyre into the northern area, whence these are likely to be transported further north by the North Equatorial Current.

REFERENCES

1. Vonder Haar, T. H. and Oort, A. H. New estimate of annual poleward energy transport by northern hemisphere oceans. J. Phys. Oceanogr. (1973) 3, 169–172.

2. Stommel, H. Oceanic warming of Western Europe. Proc. Nat. Acad. Sci. USA (1979) 76, 2518–2521.

3. Bryden, H. L. Ocean heat transport. In: Time Series of Ocean Measurements. (D. Ellet, Ed.) WCP Rep, (1983) (No. 21), WMO.

4. Khlystov, N. Z. Structure and Dynamics of Tropical Atlantic Waters. Kiev: Nauk. dumka (1976), 163 p.

5. Birman, B. A. and Pozdnyakova, T. G. Climatic Characteristics of Thermal Exchange in Areas of Active Air–Sea Interaction. Moscow: Gidrometsentr (1985), 123 p.

6. Richardson, P. L. and McKee, T. K. Average seasonal variation of the ocean. J. Phys. Oceanogr. (1984) 14, 1226–1238.

7. Hellerman, S. and Rosenstein, M. Normal monthly wind stress over the world ocean with error estimates. J. Phys. Oceanogr. (1983) 13, 1093–1104.

8. Merle, J. Seasonal variability of the subsurface thermal structure in the Tropical Atlantic ocean in hydrodynamics of the equatorial ocean. (J. C. J. Nihoul, Ed.), New York: Elsevier, (1983), 31–49.

9. Efimov, V. V. Dynamics of Wave Processes in Air–Sea Boundary Layers. Kiev: Nauk. dumka (1981), 254 p.

10. Korotaev, G. K. and Chepurin, G. A. Estimates of the vertical structure of synoptic-scale wave components. Morsk. Gidrofiz. Issled. (1982) (1), 3–11.

11. Garzoli, S. L. and Katz, E. J. The forced annual reversal of the Atlantic North Equatorial Countercurrent. J. Phys. Oceanogr. (1983) 13, 2082–2090.

12. Richardson, P. L. and Reverdin, G. Seasonal cycle of velocity in the Atlantic Equatorial Countercurrent as measured by surface drifters, meters and ship drifters. J. Geophys. Res. (1987) 92 (C4), 3691–3708.

13. Zelen'ko, A. A., Mikhailova, E. N., Polonsky, A. B. and Shapiro, N. B. Modelling of the seasonal variability of the equatorial currents system. In: Problems of the Ocean Dynamics. Leningrad: Gidrometeoizdat (1984), 70–79.

Investigations of the Tropical Atlantic Ocean, pp. 123 – 132

Diagnosis of large-scale water circulation in the Tropical Atlantic Ocean test area on the basis of four-dimensional analysis

V. V. KNYSH, V. A. MOISEENKO and V. V. CHERNOV

Abstract — The paper reports on the four-dimensional analysis as applied to the study of hydrophysical field components, with the assimilation of temperature and salinity data compiled in the central and western parts of the Tropical Atlantic Ocean test area in the course of a survey conducted in summer of 1988. Some results are given of the analysis of long currents at the diagnostic stage, as well as at the final stage of data assimilation.

One of the major goals of the Soviet RAZREZY program concerned with the study of an energetically-active zone in the Tropical Atlantic Ocean is investigation of the peculiarities of long currents and of the processes of heat accumulation and transport toward middle and high latitudes. A large amount of hydrological observations were carried out from 1986 to 1988 which have allowed a better understanding of the studied phenomena. However, in view of the data being non-synchronously collected and their corruption with noise, an adequate identification of temporal and spatial peculiarities of the large-scale circulation remains a fairly difficult task. In this context, four-dimensional analysis of the oceanic parameters based on the dynamico-stochastic modelling (DSM) of currents [1] may prove to be capable of providing the necessary solution.

To construct a numerical algorithm for assimilation of the observed data, a quasigeostrophic dynamico-stochastic model of currents was applied [2]. Its application seems reasonable in terms of updating the technology and practical use of the four-dimensional analysis by ship-mounted computers with the purpose of obtaining the characteristics of long currents in the ocean through the processing of a large amount of temperature and salinity data.

Differential equations and diagnostic relations in the DSM's hydrodynamic part includes an equation for an adynamical ocean level, an equation for heat and salinity diffusion, and relations for horizontal current velocity components, including drift components [3]. To determine the vertical current velocity, formulas in ref. [4] were used. The density of seawater is found by the nonlinear equation of state [3].

UDK 551.465.519.24

The model's statistical part is discussed in ref. [2]. This part includes differential equations and relations for calculating correlational functions of errors in the estimations of T and S, relations for calculating weight coefficients, as well as the formulae for correcting the calculated values of T and S in the course of data input.

A large-scale hydrological survey of the test area in the Tropical Atlantic Ocean was carried out from 28 July to 8 September, 1988. It's specific feature consisted of the fact that, as distinct from the former surveys, spacing between meridional sections was twice as large and equalled 3°, and the observations in the central part of the tropical test area did not have any discontinuities in time.

The technological pattern of the four-dimensional analysis of hydrophysical fields includes the following stages: preliminary processing of observations; derivation of the initial T, S-fields; computation of the autocorrelation functions of the T, S-fields; calculation of the hydrophysical parameters with a subsequent data assimilation; treatment of the results provided by the four-dimensional analysis; and computation of the secondary hydrophysical characteristics.

The preliminary processing of the observations was as follows. Expansion of the vertical T, S-profiles using empirical orthogonal functions (EOF) was performed for each hydrographic station. Twenty one levels were considered: 0, 10, 20, 35, 50, 75, ..., 200, 250, 300, 400, ..., and 1100 m. At each of these levels, the T, S-values were obtained through averaging the measured data within the intervals whose boundaries correspond to the middle range between the adjacent levels.

The estimates of T, S-expansion into series by the empirical orthogonal function recovered from five surveys carried out in 1986–1988 are listed in Table 1: λ_i is the dispersion of the averaged expansion coefficients and d_i are contributions to the total dispersion (in %) described by a sum of the i number of the expansion terms. It is seen that the expansions display a satisfactory convergence: in the case of temperature the initial 3–7 terms account for approximately 90% of the dispersion; in the case of salinity the same share of dispersion is described by the initial 3 or 4 terms of expansion. Summer and summer–autumn observational periods are specific, because 3–5 terms must be considered to describe approximately 90% of the temperature dispersion.

As the initial eigenvectors of the matrix for the T, S-disturbance correlation characterize their large-scale features, we considered four expansion terms using the data listed in Table 1 describing about 90% of the T, S-dispersion pertaining to the summertime survey of 1988. The T, S-fields filtered out in this fashion were used for construction of initial fields and data assimilation.

The initial T, S-fields at all calculational levels of the numerical model up to the depth of 1100 m were derived through optimal interpolation of the values filtered by the empirical orthogonal function to the calculated grid joints. At 1400, 2000, 3000, and 4000 m mean seasonal climatic fields were applied [5] corresponding to the season of survey. The values of the adynamical level at the initial moment of time were assumed zero.

The filtered out temperature and salinity fields were used as original material for computing their autocorrelation functions. Statistical information on the structure of the hydrophysical fields is needed for both the optimal interpolation and correction of the error correlation functions of temperature and salinity estimations, derived in the course of four-dimensional analysis.

Table 1.
Estimation of the temperature and salinity data expansion into series by the empirical orthogonal functions

Survey		Winter–spring of 1987		Spring of 1986		S u m m e r				Summer–autumn of 1986	
						1987		1988			
		λ_i	d_i	λ_i	d_i	λ_i	d_i	λ_i	d_i	λ_i	d_i
1	T	62.79	73.5	52.48	67.9	52.66	73.7	48.7	77.95	46.10	68.8
	S	0.56	53.5	0.62	42.5	1.80	66.3	3.62	74.78	1.65	53.3
2	T	3.31	77.4	9.99	80.8	7.49	84.2	7.38	89.76	10.05	83.8
	S	0.23	75.9	0.40	70.0	0.50	84.7	0.6	87.27	0.81	79.4
3	T	3.23	81.1	2.15	83.6	1.76	86.6	2.5	93.76	5.16	91.5
	S	0.10	85.5	0.22	85.3	0.14	89.9	0.31	93.77	0.28	88.4
4	T	2.33	83.9	2.14	86.3	1.42	88.6	1.37	95.95	2.44	95.1
	S	0.05	90.7	0.08	90.7	0.12	94.3	0.12	96.18	0.16	93.7
5	T	2.25	86.5	1.62	88.4	1.19	90.3	0.87	97.35	1.30	97.1
	S	0.03	93.7	0.05	94.3	0.05	96.1	0.08	97.78	0.06	95.7
6	T	1.99	8.88	1.56	90.5	0.93	91.6	0.5	98.15	0.57	97.9
	S	0.02	95.8	0.02	95.9	0.03	97.3	0.03	98.47	0.05	97.3
7	T	1.66	90.8	1.05	91.8	0.87	92.8	0.4	98.83	0.48	98.6
	S	0.01	97.2	0.01	97.0	0.02	98.2	0.02	98.96	0.03	98.2

To verify the isotropic condition of the T, S-fields, the latters' autocorrelation functions were determined for various directions with the spacing being 10°. We used the data from a survey conducted during summer 1986 as being statistically better supported compared with the summer survey in 1988. In calculating correlational functions for various directions, spatial discreteness at all considered levels was assumed equivalent to a minimal range between stations occupied in the test area, i.e. 0.5°.

Figure 1 shows the intervals of the temperature field spatial correlation. As the left-hand half-plane is symmetrical to the right-hand one, it is shown in the picture. The stations are indicated by points. The spatial statistical structure of the T, S-fields proved nonisotropic with respect to the autocorrelation function, with anisotropy being well marked along the equator. This peculiarity is observed at all levels. The length of the spatial correlation interval

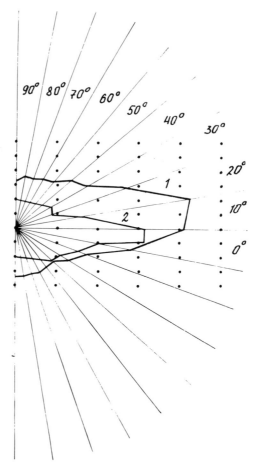

Figure 1. Intervals of the temperature field spatial correlation by a value of 0.2 m along various directions for the depth of 50 m (1) and 800 m (2).

modifies with direction by the law of ellipse. As is readily visualized in Fig. 1, the major half-axis of the correlation ellipse is slightly tilted with respect to the zonal direction. In order to consider anisotropy of the T, S-fields in the algorithm applied in four-dimensional analysis, an 'ellipse' of correlation by the functional dependence $R(x, y) = \exp[-\alpha(x - x')^2 - \beta(y - y')^2]$ was introduced, where α and β are the coefficients corresponding to the approximation by the latitudinal and longitudinal directions, respectively. The values of coefficients α and β were determined through by the least square method. In approximating the autocorrelational functions a slant of the ellipse's major half-axis was neglected.

The fields of the level sea surface, temperature, and salinity were found through numerically solving the model's differential equations. Equation difference approximation had the first order of accuracy (the directed difference method). The Coriolis parameter value within 2°30′–5°N was assumed constant and equal to its value at 5°N. A step along the x-axis was equal to 1° and along the y-axis to 30′; 16 levels were considered in the vertical: 0, 20, 50, 100, 150, 200, 300, 400, 500, 700, 900, 1400, 2000, 3000, and 4000 m. The time step was equal to a pendulum day. The turbulent exchange coefficients were as follows: $\nu = 100$, $\varkappa = 1$, and $\varkappa_t = 5 \cdot 10^7$ cm^{-2}s^{-1}. We used in computations mean seasonal climatic fields of the tangential wind stress [6]. The hydrophysical parameters were calculated for the period of survey with assimilation of the filtered T, S-observations aggregated into groups, each being 3 days long. The errors in the T, S estimates were determined by computing the difference between the filtered estimates and the computed ones. The control of assimilated temperature and salinity values was performed through estimating a magnitude of the prediction error. If these quantities were larger than 7 °C and 2.5‰, respectively, then such temperature and salinity values were ignored. Correction of the calculated T, S values, as well as the computation and correction of error dispersion in the estimates of T and S at the moment of data input was performed in the manner described in ref. [2].

Computation of the observational data being finalized, we have acquired synchronous fields of ζ, T, S, u, v, and w, streamfunction ψ at different levels, and error dispersions in the estimates of T and S derived at the final step of computation with assimilation of the observations.

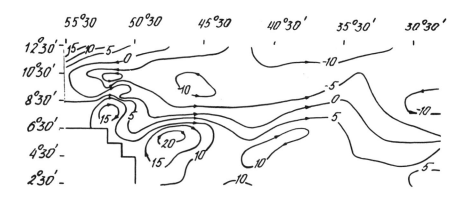

Figure 2. The diagnostic map of sea surface level (in cm) computed from filtered temperature and salinity data.

Analysis of the diagnostic sea surface level map computed from the initial temperature and salinity data (Fig. 2) shows that an intensive east-oriented stream flow identified as the North Equatorial Countercurrent (NECC), constitutes a basic element of large-scale circulation in the studied area. As the

surface gradient currents indicate, this flow is essentially quasistationary. It is
seen that the NECC is generated as result of the Guiana Current's reversal to
the east near 50–54°W and 6–8°N. An anticyclonic vortical formation with a
maximum 24.3-cm elevation of the sea surface is located there. North of this
area, the NECC is fed by the water from the North Equatorial Current. The
NECC meandering is readily seen on the sea surface map.

North of the NECC, an area of low values of sea surface level is located, and to
the south an area of higher values is located. The largest drop in the sea surface
level is 38 cm. Two cyclonic gyres are traceable in the area of low sea surface
level: one gyre occurs between 31°30′ and 40°30′W, and the other one between
42°30′ and 46°30′W. A cyclonic meander, with the center at 45°30′W and 5°N,
and an extensive anticyclonic formation between 33°30′ and 43°30′W and 2°30′
and 6°N stand out south of the NECC. The diagnostic map of the sea surface
level is constructed with the bottom topography, coastal configuration, climatic
fields of the wind stress, temperature, and salinity below 1100 m taken into
account. A well-marked reversal of the Guiana Current to the east is readily
visualized on this map, as compared with the map of dynamical topography
presented in Fig. 29 in ref. [7]. It is of interest to analyse the diagnostic currents
structure at depth. To this end, with the current assumed nondivergent, the
streamfunction $\psi(x, y, z)$ was determined at various levels from the derived
horizontal components of the current's vector velocity. The assumption on
the nondivergent current velocity field at each level is admissible, as it is
applied only at the stage of graphic data presentation. The streamfunction
was determined through solving numerically the Poisson equation

$$\Delta\psi = f(u, v).$$

The streamfunction derivative normal to the boundary at the grid boundary
points was assumed equal to the value of the respective current velocity component at the points located in the grid's marginal area. With such boundary
conditions and approximation of the streamfunction secondary derivatives by
central difference relations, a determinant in the derived system of algebraic
equations is nonzero and the latter system has a single solution.

Analysis of the diagnostic streamfunction charts in terms of depth (for the
sake of briefness, streamfunction in Fig. 3 is given for a depth of 50 m) indicates
that the NECC tends to shift south with depth. At a depth of 20 and 50 m
the NECC is observed to turn to the north in the southeastern part of the
investigated area and as the current travels eastward it becomes larger in width.
As the computations point out, the NECC is basically concentrated in the layer
between 0 and 150 m.

In the area of the NECC's genesis, its main stream is located at a 20–50 m
depth, with the largest value of the modulus of the current vector velocity
being 38 cm s^{-1} at a 20 m level. East of 39°30′W, the NECC main stream
outcrops. Its maximum velocity along the 35°30′W section at 6°30′N amounts
to 53 cm s^{-1}.

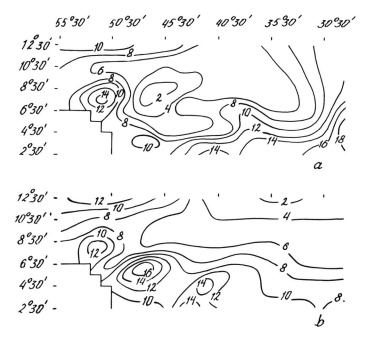

Figure 3. Streamfunction spatial distribution $(\text{cm}\,\text{s}^{-2})$ at a 50 m depth: (a) the diagnostic calculation; (b) calculations with temperature and salinity data assimilation (the final moment).

Aside from the CTD-soundings, measurements of current velocity profiles were carried out using a hydrophysical profiling probe [7]. It seems appropriate to intercompare the observed current velocity profiles with the model ones (Fig. 4). For comparison, we chose stations occupied along 32°30′W within the NECC area. It is seen that the computed velocities, as expected, gradually alter with depth. The instrument-observed current velocity profiles exhibit, along with the large-scale components, high-frequency oscillations. The absolute values of the observed velocities prove larger than the model-provided ones, and the discrepancy grows with depth. Directions of the calculated and observed current vector velocities in the upper 100 m layer being roughly identical at st. 6555, a discrepancy between the vector current velocity at st. 6553 and 6557 proves appreciable. Therefore, a vector of the instrumentally-derived current velocity at the sea surface is directed northwestward, whilst the model vector is directed eastnortheastward.

It is seen from the map of the sea surface level, constructed as of the conclusive moment of computation involving an assimilation of filtered temperature and salinity data (Fig. 5) that the sea surface level field is slightly smoothed, compared with the diagnostically computed ones. Obviously, this is explained by the time-consumung model simulations (45 days).

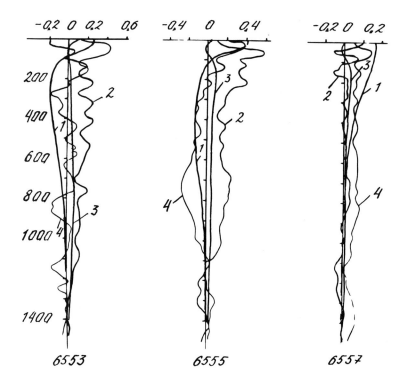

Figure 4. Zonal (1, 2) and meridional (3, 4) profiles of the current velocity components. Curves 1 and 3 are plotted using the diagnostic model data; curves 2 and 4 are based on field measurements.

The level surface in Fig. 5 is qualitatively different from that shown in Fig. 2 and from the dynamical topography of the oceanic surface (see Fig. 2a in ref. [7]). Better marked are anticyclonic gyres in the southwestern part of the studied area and in the region between 37°30′ and 42°30′W and 3°30′ and 6°30′N. These are also well pronounced in the streamfunction field (Fig. 3b).

These results were obtained because coastal geometry and bottom relief were considered by model. The NECC meanders are weakly pronounced in Fig. 5. The largest drop of the sea surface level equals 36 cm.

Two surveys were carried out during summer–autumn and summer of 1986 and 1987, respectively. Comparison of the results provided by four-dimensional analysis of the observations collected from 1986 through 1988 [2, 6, 7, 8] allows inferences. All simulations yield an intensive stream of the NECC propagating from west to east. If bifurcates in the east into two branches, i.e. the northeasterly flow and the southeasterly one. In sea surface level, charts a cyclonic feature is traceable in the western part of the studied area where the level is

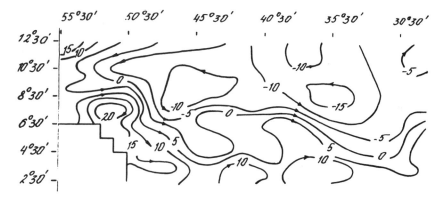

Figure 5. The map of sea surface level (cm) related to the final moment of computations with involvement of the observed temperature and salinity data.

low and an anticyclonic formation is visualized in the area of higher elevation of the sea surface level. A cyclonic eddy is observed in the area of the Guiana depression and an anticyclonic one occurs in the area of the Guiana Current's reversal to the east. An occurrence of these two quasistationary eddy formations is, apparently, related to NECC intensity and peculiarities of the bottom topography in that region.

The distinctive feature of the calculations involving the data from the summertime survey in 1988 consists in that one of the NECC meanders was located considerably further to south than during the summer–autumn period of 1986 and the summer of 1987.

Some conclusions seem appropriate. Temperature and salinity fields, filtered by presenting these by the first 4 or 5 terms of the expansion by the empirical orthogonal functions, can be used for identification of large oceanic currents by applying dynamico-stochastic modelling to four-dimensional analysis of the hydrophysical fields.

Anisotropy of the temperature and salinity fields in zonal and meridional directions of the Tropical Atlantic Ocean test area points out the existence of major differences between large-scale oceanic phenomena in these directions. In summer, large-scale circulation in the active ocean layer is characterized by the presence of flows, which are predominantly zonally-oriented.

The NECC, which becomes intensive during the summer–autumn period, may meander in the southern direction as far as 2°N.

Quasi-stationary gyres in the western Tropical Atlantic Ocean, i.e. a cyclonic gyre in the area of low surface level and an anticyclonic one in the area of high surface level evolve, apparently, by virtue of the combined effect of the bottom topography and intensification of the NECC during summer period.

REFERENCES

1. Knysh, V. V. and Timchenko, R. E. The dynamico-stochastic approach to the analysis of the variability of density and current fields in the hydrophysical test area. In: *Synoptic Eddies in the Ocean*. Kiev: Nauk. dumka (1980), 223–246.

2. Knysh, V. V., Moiseenko, V. A. and Chernov, V. V. Some results from four-dimensional analysis of hydrophysical fields in the Tropical Atlantic test area. *Izv. AN SSSR. Fizika atm. okeana* (1988) **24**, 744–752.

3. Sarkisyan, A. S. and Demin, Yu. L. (Eds). *Methods and Results of Calculating Water Circulation in the World's Ocean*. Leningrad: Gidrometeoizdat (1986), 8–10.

4. Semenov, V. V. To the calculations of vertical motions in numerical models of oceanic circulation. *Okeanologia* (1981) **21**, 433–434.

5. Levitus, S. and Oort, A. U. Global analysis of oceanographic data. *Bull. Am. Met. Soc.* (1977) **58**, 1110–1184.

6. Hellerman, S. An updated estimate of the wind stress of the World ocean. *Month. Weath. Rev.* (1986) **96**, 63–74.

7. Report on Cruise 36 of R/V *Akademik Vernadsky*. Sevastopol: MHI (1987) **1**, part 2.

8. Report on Cruise 37 of R/V *Akademik Vernadsky*. Sevastopol: MHI (1988) **1**, part 2.

Investigations of the Tropical
Atlantic Ocean, pp. 133 – 148

Modelling of the seasonal variability in the Tropical Atlantic Ocean

E. N. MIKHAILOVA, I. M. SEMENYUK and N. B. SHAPIRO

Abstract — Discussed are the results of numerical modelling of the seasonal variability of circulation, temperature and salinity fields in the Tropical Atlantic Ocean in the framework of a multilevel nonlinear baroclinic model which considers the real geometry of the basin, the bottom topography, and a number of external factors (surface wind stress, sea surface temperature, and salinity).

Simulation of the seasonal course of circulation and thermohydrodynamical fields in the tropical region of the world's ocean remains one of the challenging problems of numerical modelling of ocean dynamics [1]. Oceanographic research conducted in recent years [2–6] indicate that a decisive role in the seasonal variability of the tropical ocean fields belongs to seasonal wind fluctuations. Moreover, for qualitative simulation of basic features of the nonstationary Tropical Atlantic Ocean circulation by any model (linear, nonlinear, layered, continuously-stratified), it suffices to consider the most characteristic properties of the wind field, in particular, the calm zone located between the trade winds, in other words, the Intratropical Convergence Zone (ITCZ) and its seasonal displacement in the meridional direction. Even with the schematically–prescribed wind field involving temporally jumplike [3] or smooth [6] shifting of the ITCZ, the models describe the principal peculiarities of the restructuring of the Equatorials Currents, the North Equatorial Countercurrent (NECC) and the properties of the related temperature field. Among these we may point out, for example, intensification of the NECC, its shifting to the north, and emergence of a frontal zone in the temperature field during the ITCZ's movement in a northerly direction from winter to summer.

Intercomparison of different models shows that nonlinear baroclinic models based on primitive equations using conservative numerical schemes [5, 6] seem to be most suitable for both the qualitative simulation and the quantitative one of the seasonal course of the ocean's hydrodynamical fields. In ref. [6], seasonal variability of the Tropical Atlantic is examined without considering its interaction with the middle latitudes and neglecting the effect of bottom topography (in the 'baroclinic approximation') with the model external factors prescribed. The problem is handled in a more complete fashion in ref. [5]:

UDK 551.465

the model considers the influence of the middle latitudes, the bottom topography, to describe the upper ocean layer more accurately the real external factors were prescribed and variable vertical turbulent exchange coefficients (functions of the Richardson number) applied. Numerical simulation in ref. [6] involved a conservative monotonic scheme of the first order of accuracy in space and time, and the notorious Bryan's scheme of the second order of accuracy [7] was used in ref. [5]. Intercomparison of the indicated numerical schemes in the context of the problem of oceanic fields adaptation in the Tropical Atlantic Ocean [8] shows that yielding, on the whole, sufficiently similar results, each of these schemes has its own certain advantages and disadvantages. The monotonic scheme of the first order of accuracy seems more economical, though schematic viscosity and diffusion, inherent to it, may lead to a smoothing of the solution (particularly in the areas of jet currents). In nonmonotonic schemes of the second order of accuracy, the obtained results are distorted due to the computational dispersion evolving in the form of two-step waves, this being the case when spatial resolution is insufficient. In order to inhibit it, an additional dissipation is necessarily introduced. An economical conservative scheme suggested in ref. [8] uses, to a certain extent, advantages inherent to both schemes. Only a dosed schematic viscosity is introduced there at the segments, where two-step waves are likely to occur in space. This scheme was applied to the problem of diagnosis and adaptation of Tropical Atlantic Ocean hydrothermodynamical fields, and proved to be more efficient, as compared to the 1st order conservative scheme with directed differences.

The purpose of this paper is modelling of seasonal variability of the Tropical Atlantic Ocean in the most complete fashion using the indicated numerical methods with the exchange coefficients being relatively small. The model is supposed to consider salinity (as distinct from other simulations), the real coastal configuration and bottom topography, the real (climatic) seasonally-variable wind fields [9], and the prescribed sea surface temperature and salinity fields [10].

The original system of equations, the boundary and initial conditions will read:

$$\frac{du}{dt} - \beta yv = g\frac{\partial \xi}{\partial x} - \int_0^z \frac{\partial \sigma}{\partial x}dz + A\frac{\partial^2 u}{\partial z^2} + A_l\Delta u,$$

$$\frac{dv}{dt} + \beta yu = g\frac{\partial \xi}{\partial y} - \int_0^z \frac{\partial \sigma}{\partial y}dz + A\frac{\partial^2 v}{\partial z^2} + A_l\Delta v; \tag{1}$$

$$\frac{\partial u}{\partial x} + \frac{\partial v}{\partial y} + \frac{\partial w}{\partial z} = 0; \tag{2}$$

$$\frac{dT}{dt} = \mu\frac{\partial^2 T}{\partial z^2} + \mu_l\Delta T; \tag{3}$$

$$\frac{dS}{dt} = \mu\frac{\partial^2 S}{\partial z^2} + \mu_l\Delta S; \tag{4}$$

$$\sigma = \sigma(T, S); \tag{5}$$

at the sea surface at $z = 0$

$$A\frac{\partial u}{\partial z} = -\tau^x(x, y, t), \qquad A\frac{\partial v}{\partial z} = -\tau^y(x, y, t), w = 0,$$

$$T = T_0(x, y, t), \qquad S = S_0(x, y, t). \tag{6}$$

at the bottom at $z = H(x, y)$

$$A\frac{\partial u}{\partial z} = 0, \qquad A\frac{\partial v}{\partial z} = 0, \qquad w = u\frac{\partial H}{\partial x} + v\frac{\partial H}{\partial y},$$

$$\mu\frac{\partial T}{\partial z} = 0, \qquad \mu\frac{\partial S}{\partial z} = 0; \tag{7}$$

at the boundaries of the basin $(x, y) \in l$

$$u = v = 0, \qquad \frac{\partial T}{\partial n} = \frac{\partial S}{\partial n} = 0; \tag{8}$$

at $t = 0$

$$u = u^0(x, y, z), \quad v = v^0(x, y, z), \quad T = T^0(x, y, z), \quad S = S^0(x, y, z). \tag{9}$$

In relations (1)–(9), u, v, and w are the current velocity components along the x-, y-, and z-axes directed, respectively, eastward (along the equator), northward, and vertically downward, $\sigma = g(\rho - \rho_0)/\rho_0$ is the buoyancy; $g = 980 \text{ cm s}^{-2}$; and $\rho_0 = 1.027$ is the abyssal mean density. Equation of state (5) is taken in a nonlinear form with the quadratic terms considered [11], specifically, $\rho = a_0 - a_1 T - a_2 T^2 + a_3 S - a_4 TS$, $a_0 = \text{const}$, $a_1 = 35 \times 10^{-7}$, $a_2 = 8.02 \times 10^{-4}$, $a_3 = 2 \times 10^{-6}$, and $a_4 = 469 \times 10^{-8}$; temperature is expressed in °C and salinity in ‰; ξ is the sea surface level, and τ^x and τ^y are the wind stress components. Vertical (A, μ) and horizontal (A_l, μ_l) exchange coefficients are assumed constant.

$$\frac{d\varphi}{dt} \equiv \frac{\partial\varphi}{\partial t} + \frac{\partial v\varphi}{\partial y} + \frac{\partial w\varphi}{\partial z},$$

$\partial/\partial n$ is the derivative normal to the boundary.

Let us describe the finite–difference scheme of solving problem (1)–(9). The original equations are approximated using the box-method applied to grids which are shifted horizontally (grid B) and vertically (the nonuniform grid); the vertical current velocity is determined at the sea surface, at the bottom, and at depths being at a central position between the two levels where the other quantities are being determined. Time integration is performed using a two-layer semi-implicit scheme of the first order of accuracy (n is the number of a time step):

$$\frac{u^{n+1} - u^n}{\Delta t} - \beta y \frac{v^{n+1} + v^n}{2} + \left\langle \frac{\partial u^2}{\partial x} \right\rangle^n + \left\langle \frac{\partial uv}{\partial y} \right\rangle^n + \left\langle \frac{\partial u^{n+1} w^n}{\partial z} \right\rangle$$

$$- \left\langle A \frac{\partial^2 u}{\partial z^2} \right\rangle^{n+1} - \left\langle A_l \Delta u \right\rangle^n + \left\langle \int_0^z \frac{\partial \sigma}{\partial x} \, dz \right\rangle^{n+1} - \left\langle g \frac{\partial \xi}{\partial x} \right\rangle^{n+1} = 0,$$

$$\frac{v^{n+1} - v^n}{\Delta t} + \beta y \frac{u^{n+1} + u^n}{2} + \left\langle \frac{\partial uv}{\partial x} \right\rangle^n + \left\langle \frac{\partial v^2}{\partial y} \right\rangle^n$$

$$+ \left\langle \frac{\partial v^{n+1} w^n}{\partial z} \right\rangle - \left\langle A \frac{\partial^2 v}{\partial z^2} \right\rangle^{n+1} - \left\langle A_l \Delta v \right\rangle^n \qquad (10)$$

$$+ \left\langle \int_0^z \frac{\partial \sigma}{\partial y} \, dz \right\rangle^{n+1} - \left\langle g \frac{\partial \xi}{\partial y} \right\rangle^{n+1} = 0;$$

$$\left\langle \frac{\partial u}{\partial x} \right\rangle^n - \left\langle \frac{\partial v}{\partial y} \right\rangle^n + \left\langle \frac{\partial w}{\partial z} \right\rangle^n = 0; \qquad (11)$$

$$\frac{T^{n+1} - T^n}{\Delta t} + \left\langle \frac{\partial uT}{\partial x} \right\rangle^n + \left\langle \frac{\partial vT}{\partial y} \right\rangle^n + \left\langle \frac{\partial w^n T^{n+1}}{\partial z} \right\rangle$$

$$- \left\langle \mu \frac{\partial^2 T}{\partial z^2} \right\rangle^{n+1} - \left\langle \mu_l \Delta T \right\rangle^n = 0; \qquad (12)$$

$$\frac{S^{n+1} - S^n}{\Delta t} + \left\langle \frac{\partial uS}{\partial x} \right\rangle^n + \left\langle \frac{\partial vS}{\partial y} \right\rangle^n + \left\langle \frac{\partial w^n S^{n+1}}{\partial z} \right\rangle$$

$$- \left\langle \mu \frac{\partial^2 S}{\partial z^2} \right\rangle^{n+1} - \left\langle \mu_l \Delta S \right\rangle^n = 0; \qquad (13)$$

$$\sigma^n = \sigma(T^n, \, S^n). \qquad (14)$$

The terms between the angular brackets are difference analogues to the respective spatial differential operators. The vertical velocity is determined from the discontinuity equation individually to compute the current velocity components, temperature and salinity, as these are determined at different points. Spatial approximation of the diffusion operators is performed in a common way. Operators with the horizontal advection are presented as differences between the flows across the opposite box edges, which are written with the second or first order of accuracy depending on the sign of velocity and the gradient of the considered substance (the updated scheme of directed differences) [8]. Directed differences are applied merely when a sign of the current velocity component coincides with that of the respective substance gradient component, otherwise the schemes of the second order of accuracy are used. For instance,

$$\left\langle \frac{\partial u\varphi}{\partial x} \right\rangle_i = \frac{1}{\Delta x}\left(u_{i+1/2}\cdot \widetilde{\varphi}_{i+1/2} - u_{i-1/2}\cdot \widetilde{\varphi}_{i-1/2} \right),$$

where

$$\widetilde{\varphi}_{i+0.5} = \begin{cases} \varphi_{i+1}, & \text{if } u_{i+0.5} \leqslant 0 \text{ and } |\varphi|_{i+1} \leqslant |\varphi|_i; \\ \varphi_{i-1}, & \text{if } u_{i+0.5} > 0 \text{ and } |\varphi|_{i+1} > |\varphi|_i; \\ (\varphi_{i+1} + \varphi_i)/2 & \text{in the other cases.} \end{cases}$$

With such approximation, like in the case involving the common scheme of the first order of accuracy, the main physical laws are not violated at the level of interaction of separate boxes. In particular, the heat is transported only from a warm box to a cold one and momentum is transported from a box with large impulse to a box with small impulse, whereas in the scheme of the second order of accuracy at $\widetilde{\varphi}_{i+0.5} \equiv (\varphi_{i+1} + \varphi_i)/2$, heat, for instance, may be transported from a cold box to a warm one.

In computations of the vertical velocity, temperature, and salinity profiles by the implicit scheme, vertical advection is presented via directed differences, and the quantities are then recalculated following the principle of correction of flows [12] with schematic viscosity and diffusion being subtracted. To integrate the equations of motion and the temperature and salinity equations, the methods of matrix runs and conventional runs are applied.

For level computations, and equation for the integral streamfunction is used. It is derived directly from the vertically summarized difference equations for motion and continuity by eliminating level inclinations from the latter and is solved by the method of upper relaxation.

Computations were performed for the region, which approximates the Tropical Atlantic Ocean region confined between 10°S and 15°N (with the Caribbean Sea ignored). The bottom relief is approximated by steps, which coincide with the box edges closest to the bottom, i.e. the levels at which vertical velocity is determined. In the horizontal, a grid with 1° spacing was used. Horizontal current velocity components, temperature, and salinity were computed for 16 levels: 5, 20, 30, 50, 75, 100, 150, 200, 300, 500, 700, 1000, 1500, 2000, 3000, and 4000 m. The greatest depth of the ocean was 5000 m, and a minimum one 125 m. Exchange coefficients were as follows: $A = 2\mu = 2 \text{ cm}^2\,\text{s}^{-1}$ and $A_L = \mu_L = 10^7 \text{ cm}^2\,\text{s}^{-1}$. The time step was 6 h.

Seasonally-variable fields of wind, sea surface temperature, and salinity were imposed in the following manner: mean monthly climatic wind data series were expanded into harmonics and the annual and semi-annual harmonics used. The same procedure was performed with the temperature and salinity data for the four seasons. In the capacity of initial fields, we used temperature and salinity data collected in summer [10] and the current velocity fields recovered from these by the diagnostic method [8]. For the initial 18 months, a scheme of the first order of accuracy was used for calculations. Hydrothermodynamical fields obtained in the course of this operation were used then as the initial ones for computing the annual course by different schemes. Thus we extended for another year computations of the seasonal cycle using the scheme of the first-order of accuracy, the updated scheme and the scheme of the second-order accuracy [8], which is similar to Bryan's scheme but involves larger coefficient values. In the present study, we focus on the results obtained using the updated scheme of directed differences.

Let us consider results derived from the numerical experiment. Figure 1 shows the calculated vector fields of the horizontal current velocity compo-

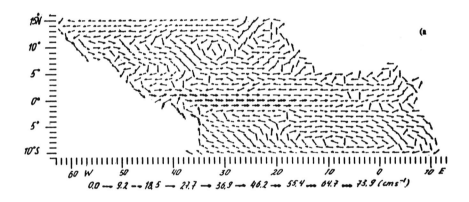

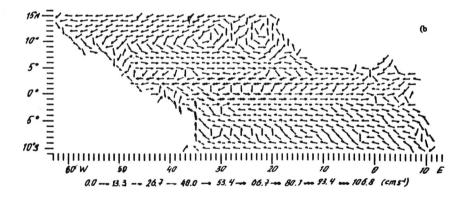

Figure 1. Current velocity at a depth of 50 m at the end of winter (a), spring (b), summer (c), and autumn (d).

nent at 50 m depth observed at the end of winter (a), spring (b), summer (c), and autumn (d). At the end of winter, when the ITCZ is in its southern most position reaching the equator in the western Tropical Atlantic Ocean, intensive currents virtually do not occur at this depth outside the equatorial region. The Equatorial Undercurrent (the Lomonosoff Current), propagating in the form of a jet stream from the western shore nearly to the eastern shore, its speed reaching 65–75 cm s^{-1} in the area between 35 and 20°W, is emphatically pronounced there. A weak easterly flow is visualized north of the equator. As refs. [3, 6] claim, this flow seems to be the degradated NECC which was travelling there in autumn and the newly-evolving NECC in the

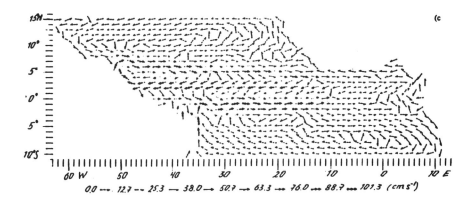

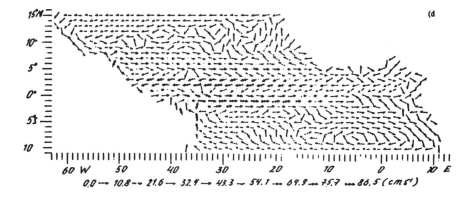

Figure 1. Continued.

calm area confined between the trade winds converges with the Equatorial Undercurrent. The easterly current mentioned above is a continuation of the southeasterly current propagating along the western boundary from the north. This southeasterly current bifurcates near 40–45°W and one of its branches reverses and flows into the Equatorial Undercurrent. With the NECC not being a jet stream, the eddies occur at its northern periphery whose velocities reach 18 cm s^{-1} at most. A weak westerly current (the northern branch of the South Equatorial Current) is located between the NECC and the Equatorial Undercurrent, its velocity being less than 10 cm s^{-1}. On the sea surface (except for the northern part of the Gulf of Guinea), west-oriented flows prevail at this time of the year; more specifically, they are directed northwestward north of the equator, i.e. a divergence of surface currents takes place at the equator. The northwesterly current intensifies near the north coast of South America in the area bounded by 1 and 3°N (the Guiana Current), where velocities become as large as 110 cm s^{-1}. Note that at 50 m depth the current flows in the opposite direction (the Antilles–Guiana Countercurrent).

The currents' structure at 75- and 100 m depths is similar to that at 50 m. At 200 m the picture drastically changes. The Equatorial Undercurrent virtually vanishes and only in the western part of the ocean as far as 30°W a relatively intensive easterly flow, its largest velocity being 20 cm s^{-1}, is observed at 1–2°N. This flow represents the southern periphery of a large cyclonic eddy. At larger depths (500–2000 m levels), the currents are rather weak (~ 5 cm s^{-1}), forming a system of nearly-zonal flows of different directions (near the equator these are oriented westward) with relatively large eddies occurring near the northern and southern boundaries of the basin.

At the end of spring, the ITCZ shifts to the north to occupy its middle position at 30°W and 5°N. A distinct easterly jet current of moderate intensity (the NECC) evolves then at a depth of 50 m in the vicinity of the ITCZ; the current velocity is largest near its southern periphery attaining 50 cm s^{-1} at 25–20°W. The current virtually reaches the eastern coast of the Atlantic bifurcating near the African coast: one branch of the current turns to the north, and the other one penetrates into the Gulf of Guinea. Due to intensification of the southeasterly trade wind [1], the Equatorial Undercurrent shifts below the equator; it is observed to be most intensive in the western section of the ocean where its velocity reaches 80 cm s^{-1}. An intensive westerly jet current flows between the two countercurrents which is the northern branch of the South Equatorial Current (the southern branch is not jet-like). The current crosses the entire Atlantic Ocean being most intensive in the central part where its velocity attains 80 cm s^{-1}. In approaching the South Americans coast, this westerly current bifurcates, its branches feeding the Equatorial Undercurrent and the NECC, respectively. Aside from this source, the NECC is fed by waters coming from the north and the Equatorial Undercurrent receives waters from the south which are brought primarily by the coastal stream. It is appropriate to note the presence of several large eddies north of the NECC in the eastern section of the basin; an anticyclonic eddy between the NECC and the

South Equatorial Current near the western ocean border, and oscillations of the Equatorial Undercurrent axis in its western section. The pattern of ocean surface currents drastically changes at this time of the year, mainly due to the reversal of the northwesterly current to the east and genesis of a relatively intensive (up to 100 cm s^{-1}) NECC in the eastern section of the ocean. The Guiana Current's velocity then becomes as large as 160 cm s^{-1}. Note that excessively large velocities of the computed surface currents, particularly of the near-equatorial west-oriented ones, occur because of the small vertical viscosity coefficients imposed for the surface ocean layer. At a depth of 75, 100, and 200 m the pattern of currents is similar to that at 50 m, although the intensity is smaller. At larger depths (500, 2000 m) the currents become weak. Compared with the winter season, the near-equatorial current travelling at a 500 m depth is directed eastward.

At the end of summer the ITCZ reaches its northernmost position at 30°W and 7–9°N, where the wind velocity zonal component alternates its direction from east to westward. As is seen in Fig. 1, all jet currents attain maximum intensity at this time of the year. Zonal currents prove to be most intensive in the western section of the basin at a depth of 50 m. Velocities of the NECC and of the South Equatorial Current's northern branch attain 90 cm s^{-1}, the Equatorial Undercurrent's velocity is as large as 100 cm s^{-1} and the Guiana Current's velocity reaches 75 cm s^{-1}. This current is a continuation of the northern branch of the South Equatorial Current. At about 7°N the current reverses and converges with the NECC. The fact that the Guiana Current penetrates relatively far to the north in order to return then south and joint the NECC points to a significant role of nonlinear effects in the hydrodynamics of the considered region. Note that the NECC is accommodating waters transported not only from the south but also from the north (the Antilles–Guiana Counter current waters). In fact, the NECC constitutes the general border for the zonally-stretched anticyclonic and cyclonic gyres. The Equatorial Undercurrent, shifted towards 1°S, is traditionally fed by waters transported from the north and south regions. At this time of the year the most intensive are currents on the sea surface, with the trans-Atlantic North Equatorial Counter current being distinctly pronounced. The structure of currents at 75 and 100 m is similar to that at 50 m. At a 200 m level the considered zonal jet currents are marked only in the western section of the Atlantic Ocean, the Guiana Current crucially degradates and the NECC is supported at this depth basically by the waters coming from the north along the western border. The northern branch of the South Equatorial Current feeds the Equatorial Undercurrent whose axis oscillates near the equator. An extensive west-oriented flow near the equator with a ∼ 10 cm s^{-1} velocity may be singled out at a depth of 500 m. At a 2000 m level, we obtain a vortical structure with relatively feeble currents (the velocity being of the order of 5 cm s^{-1}).

At the end of autumn the calm area (ITCZ), after a shift to the south, resumes its central position. Circulation is maintained at 50 m depth, though the jet currents become less intensive. A maximum velocity of the NECC (in the eastern Tropical Atlantic Ocean) amounts to 65 cm s^{-1}, that of the Equatorial Undercurrent (in the central Tropical Atlantic Ocean) and of the South Equatorial Current's northern branch (in the central Tropical Atlantic Ocean) is 85 cm s^{-1}. Vortical activity of the NECC becomes more pronounced at its northern periphery. The southeasterly current near the western ocean border, whose waters of northern origin contribute to the NECC becomes more intensive. However, the Guiana Current degradates (~ 40 cm s^{-1}). The Equatorial Undercurrent axis shifts closer to the equator, and the current proper becomes less intensive in the eastern section of the ocean. The currents' pattern at the sea surface changes and becomes similar to that evolving at the end of spring, when the NECC is pronounced only in the eastern Tropical Atlantic Ocean. At a depth of 75 and 100 m the currents' structure is exactly like the one observed at the 50 m level. At a depth of 200 m, the NECC is traceable at least in the western Tropical Atlantic Ocean, the Equatorial Undercurrent being nontraceable at all, and the northern branch of the South Equatorial Current crucially weakens. At deeper layers (500, 2000 m), a vortical structure with feeble flows is prevalent.

The numerical experiment allows the following conclusions. The NECC exists as an intensive jet current from April to November, its velocity being maximum from July to September. In contrast to the experimental results in ref. [6], the countercurrent's core shifts insignificantly in the meridional direction from season to season, but with its width changing considerably (maximum at the end of summer). The NECC is explicitly asymmetrical, specifically, a meridional bias of the current's zonal velocity component at the southern periphery of the countercurrent is much larger than at the northern one; the NECC axis is located near the counter current's southern boundary. Disappearance of the NECC during the winter–spring period is confirmed by observations and is consistent with computations conducted by other researchers [5].

In ref. [3], where the problem of oceanic response to an instantaneous displacement of the ITCZ is addressed, it was concluded that two east-oriented jets simultaneously evolve when the ITCZ shifts south. In this case, the NECC does not also move south — it attenuates and a new easterly current is generated in the southern area where the ITCZ moves. The present numerical experiment involving 'real' climatic wind field data, as well as computations in ref. [6], where the ocean's response to gradual shifting of the ITCZ is examined, fail to yield two coexisting east-oriented jets. As the presence of such flows, at least in subsurface layers, is evidenced by the field observations (the same is confirmed, for example, by the occurrence of two local minimums of the west-oriented surface currents in spring [13]), we may conclude that a more accurate description of the space–time structure of the wind field is required: climatic fields, obviously, fail to depict fast migrations of the ITCZ between the trade winds.

The northern branch of the South Equatorial Current, being a west-oriented flow, exists throughout the year but becomes intensive and jet-like only when the NECC intensifies.

The Equatorial Undercurrent is observable throughout the year. At 30°W its zonal velocity component has two local maxima: at the end of winter and at the end of summer. The current's axis migrates between the equator (the end of winter) and 2°N (summer–autumn).

The Guiana Current, a flow basically oriented northwestward, remains active near the western border of the basin north of the equator throughout the year. It is most intensive during the summer–autumn period when it encompasses the upper 200 m layer and stretches from the South Equatorial Current up to the NECC. The rest of the time it acts as a surface flow travelling from the equator up to the northern boundary of the basin.

Let us now consider variability of the thermodynamic fields. It should be reminded that distribution of these fields at the sea surface is based on observational data. The most characteristic feature of the sea surface temperature consists in the presence of an area of high temperature located in the vicinity of the ITCZ and shifting together with the latter. As is indicated in ref. [6], this feature of the sea surface temperature is related to the wind speed, which leads to the genesis and evolution of the currents convergence zone in this region of the world's ocean. In connection with this, it can be expected that temperature distribution in intermediary and deep sea layers is not governed by temperature values imposed at the sea surface but is associated with the system of the equatorial currents and the distribution of the areas of currents divergence and convergence. Figure 2 shows temperature distributions at a depth of 50 m at the end of winter (a), spring (b), summer (c), and autumn (d). It is seen here that the temperature field displays in winter some peculiarities, although less pronounced than in the other seasons, which are related to the NECC and the Equatorial Undercurrent. Among these is the frontal zone, where temperature decreases to the north being consistent with the respective rise of the thermocline, and a local minimum of temperature occurring along the equator. In addition, some other peculiarities should be indicated, i.e. the local maximum of temperature near the northwestern border of the basin at 5°N, the local minima at the eastern border in the northern area and in the Gulf of Guinea, as well as the general rise of temperature from east to west. These peculiarities remain observable up to a depth of 100 m with the equatorial minimum of temperature decreasing and are traceable only in the central Tropical Atlantic Ocean. Alternatively, a local maximum of temperature intensifies near the northwestern boundary. At a depth of 200 m, the equatorial minimum of temperature vanishes, being traceable only in the north-west, and the entire equatorial section of the ocean exhibits a local maximum of temperature (the effect of thermocline stretching). Temperature fluctuations at 500 m deep become less spectacular; the sign of the zonal temperature gradient alternates, and an area with a local minimum of temperature emerges and is now stationed in the vicinity of the northwestern border at 5°N. At 2000 m the temperature is rising again from east to west and a local minimum of temperature occurs anew near the equator.

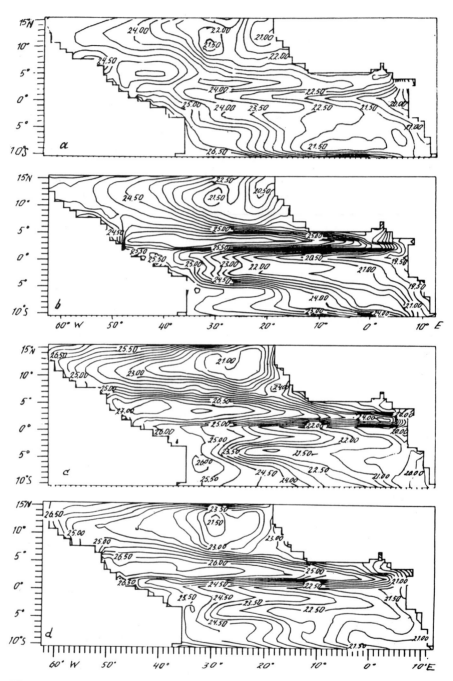

Figure 2. Temperature isolines at the end of winter (a), spring (b), summer (c), and autumn (d).

At the end of spring, when 50 m deep zonal currents intensify and the frontal zones related to the NECC and to the northern branch of the South Equatorial Current become sharply pronounced, a zonally-elongated warm water area stretches from shore to shore between these frontal zones. Temperature minimum near the equator and a local maximum south of it, coupled with the Equatorial Undercurrent's activity are maintained. In general, the temperature is higher than in winter time. Up to 100 m deep the indicated features persist at least outside the Gulf of Guinea. In the currents convergence zone, the high temperature area shifts toward the western border, where at 200 m the frontal zones are no longer observable and the local temperature extremums persist, with the local maximum being observable now only near the equator. At depths of 500 and 2000 m, the temperature distribution remains virtually the same as at the end of winter.

At the end of summer and autumn, the temperature at all ocean levels is strongly reminiscent of the temperature at the end of the spring season. Discrepancies are associated merely with the geometry (location, width, length) of frontal zones, this being readily visualized in the 50 m level case.

Intercomparison of the computed temperature field with the observations [13] provides their satisfactory agreement. However, computations yield higher temperatures in the areas of local minima at the eastern border of the ocean. This occurs, obviously, owing to the ignoring of the cold Canary Current by the model which is generated outside the considered region, i.e. the model does not take into account the influence of the middle latitudes upon the Tropical Atlantic Ocean.

Let us briefly discuss computations of the salinity field. It should be reminded that surface salinity, just like surface temperature, was imposed using observational data. The main peculiarity of the field consists of the presence and evolution of local low-salinity zones evolving due to the freshening effect of the Amazon, Niger and Congo rivers. Figure 3 exhibits salinity distributions at the 50 m level at the end of winter (a), spring (b), summer (c), and autumn (d). The freshening effects do not manifest themselves at this depth, and a system of local salinity tongues related to the equatorial currents system is visualized. Therefore, the local salinity maximum occurring in the Equatorial Undercurrent is distinctly traceable in all seasons. Besides, one can see tongues of high and low salinity related to the NECC and the northern branch of the South Equatorial Current, respectively. Seasonal variability of salinity at this depth, as seen in Fig. 3, manifests itself through an insignificant variability of the intensity of the indicated tongues of high or low salinity. At the 75 and 100 m levels, the main features of salinity distribution are roughly identical to those at 50 m deep. Below 200 m, salinity and its horizontal gradients decrease. In general, the obtained salinity distributions are consistent with the known observational data [14] although the model does not consider the inflow of saline waters into the equatorial zone from the subtropical gyre. The agreement is apparently achieved owing to the prescription of a real three-dimensional salinity field at the initial moment of time. In the course of computation, salinity is merely redistributed due to the variability of the currents field.

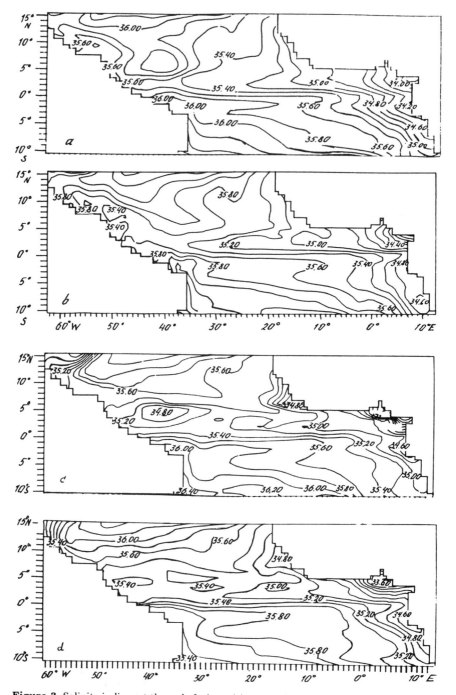

Figure 3. Salinity isolines at the end of winter (a), spring (b), summer (c), and autumn (d).

Note that in comparison to ref. [5], our computations have a poorer resolution in the vertical (the number of considered levels is approximately twice as small) and in the horizontal (the meridional step is three times as small). Besides, as has been indicated above, another condition for sea surface temperature is applied: the model takes into account neither the interaction of the tropical region with the middle latitudes nor the variability of the vertical exchange coefficients in the upper oceanic layer. Notwithstanding the dosed schematic viscosity, the computed intensity of quasizonal flows has proved similar to and in some cases even larger than in ref. [5]. Also the main features of the currents' variability and of the related temperature fields are similar. In contrast to ref. [5], we have computed in the present paper the seasonal variability of salinity in the Tropical Atlantic Ocean which agrees with the available observational data.

It may be concluded that a majority of specific features of seasonal variability in the tropical ocean are determined only by the internal dynamics of the tropical zone of the ocean, in other words, oceanic phenomena occurring in the tropical ocean region depend little on those which take place in the mid-latitude ocean. Such inference holds for the discussed problems, where in fact, only the evolution of initial fields is examined. Regretfully, we have not succeeded as yet in studying seasonal variability of the tropical ocean using a complete nonlinear model which would not be affected by the initial fields. The problem consists in the variety of technical difficulties hampering long-term computations as well as the inaccuracies in numerical simulations of the three-dimensional ocean structure involving current turbulent exchange parametrizations, which do not allow, for instance, an adequate description of the thermocline's vertical structure.

In the present study, we have restricted ourselves only to the analysis of hydrothermodynamical fields which are characteristic in general for all seasons. A fine analysis of the intraseasonal restructuring, including manifestations of wave processes in the vicinity of jet currents, and intecomparison of the data provided by different numerical schemes have not been addressed in the above discussion. Our main objective was to demonstrate that the suggested numerical model is capable of providing good results and can be applied for calculations involving the real external factors.

REFERENCES

1. Korotaev, G. K., Mikhailova, E. N. and Shapiro, N. B. *Theory of Equatorial Countercurrents in the World's Ocean.* Kiev: Nauk. dumka, (1986), 208 p.

2. Bucalacchi, A. J. and Picant, J. Seasonal variability from the model of the Tropical Atlantic Ocean. *J. Phys. Oceanogr.* (1983) **13**, 1564–1588.

3. Zelenko, A. A., Mikhailova, E. N., Polonsky, A. A. and Shapiro, N. B. *Problems of Ocean Dynamics.* Leningrad: Gidrometroizdat, (1984) 70–79.

4. Du Penhoat, Y. and Treguer, A. M. The seasonal linear response of the Tropical Atlantic Ocean. *J. Phys. Oceanogr.* (1985) **15**, 316–329.

5. Philander, S. G. H. and Pacanowski, R. C. A model of the seasonal cycle in the Tropical Atlantic Ocean. *J. Geophys. Res.* (1986) **90** (C12), 14,192–14,206.

6. Mikhailova, E. N., Semenyuk, I. M. and Shapiro, N. B. Response of the Tropical Atlantic to seasonal variations of the ITCZ. *Morsk. Gidrofiz. Zh.* (1988) **4**, 33–40.

7. Bryan, K. A numerical method for the study of the circulation of the World ocean. *J. Comput. Phys.* (1969) **4**, 347–376.

8. Mikhailova, E. N. Semenyuk, I. M. and Shapiro, N. B. Intercalibration of numerical models for the Tropical Atlantic. Sevastopol: MHI, Ukr. SSR Ac. of Sc., preprint, (1989) 25 p.

9. Hellerman, S. and Rosenstein, M. Normal monthly wind stress over the World ocean with error estimates. *J. Phys. Oceanogr.* (1983) **13**, 1093–1104.

10. Levitus, S. *Climatological Atlas of the World Ocean.* NOAA. Prof. Pap. (1982) **13**, 173 p.

11. Ibraev, R. A. Numerical modelling of large-scale hydrophysical fields in the equatorial ocean. Doctoral dissertation. Moscow: (1985), 132 p.

12. Shmakin, A. I. and Fursenko, A. A. About one monotonical difference scheme for the through calculation. *Zh. vych. matematiki mat. fiziki* (1980) **20**, 1021–1031.

13. Richardson, P. L. and McKee, T. K. Average seasonal variation of the Atlantic equatorial currents from historical ship drifts. *J. Phys. Oceanogr.* (1984) **14**, 1226–1238.

14. Merle, J. *Atlas Hydrologique Saisonnier de l'Océan Atlantique Intertropical.* Paris: Orstom (1978), 184 p.